T0260433

Triumph 350 & 500 Unit Twins
Owners Workshop Manual

by Clive Brotherwood

Models covered

(137 – 2AG9)

348cc	3TA (Twenty-One)	1957 – 1966
	T90	1962 – 1968
490cc	5TA	1958 – 1966
	T100A	1959 – 1961
	T100SS	1961 – 1970
	T100T	1966 – 1970
	T100R	1970 – 1973
	T100C	1970 – 1973

© Haynes Group Limited 2007

All rights reserved. No part of this book may be reproduced or transmitted in any form or by any means, electronic or mechanical, including photocopying, recording or by any information storage or retrieval system, without permission in writing from the copyright holder.

ISBN 978 0 85696 137 3

British Library Cataloguing in Publication Data
A catalogue record for this book is available from the British Library.

Printed in India

Haynes Group Limited
Sparkford, Yeovil, Somerset BA22 7JJ, England

Haynes North America, Inc
2801 Townsgate Road, Suite 340, Thousand Oaks, CA 91361, USA

ABCDE
FGHIJ
KLMNO
P
3

ILLEGAL COPYING

It is the policy of the Publisher to actively protect its Copyrights and Trade Marks. Legal action will be taken against anyone who unlawfully copies the cover or contents of this Manual. This includes all forms of unauthorised copying including digital, mechanical and electronic in any form. Authorisation from the Publisher will only be provided expressly and in writing. Illegal copying will also be reported to the appropriate statutory authorities.

Disclaimer

There are risks associated with automotive repairs. The ability to make repairs depends on the individual's skill, experience and proper tools. Individuals should act with due care and acknowledge and assume the risk of performing automotive repairs.

The purpose of this manual is to provide comprehensive, useful and accessible automotive repair information, to help you get the best value from your vehicle. However, this manual is not a substitute for a professional certified technician or mechanic.

This repair manual is produced by a third party and is not associated with an individual vehicle manufacturer. If there is any doubt or discrepancy between this manual and the owner's manual or the factory service manual, please refer to the factory service manual or seek assistance from a professional certified technician or mechanic.

Even though we have prepared this manual with extreme care and every attempt is made to ensure that the information in this manual is correct, neither the publisher nor the author can accept responsibility for loss, damage or injury caused by any errors in, or omissions from, the information given.

Acknowledgements

Our thanks are due to the Triumph Engineering Company Limited for their assistance. Brian Horsfall gave the necessary assistance with the overhaul and devised ingenious methods for overcoming the lack of service tools. Les Brazier took the photographs that accompany the text. Jeff Clew edited the text and John Murphy originated the layout.

Further thanks are due to Bryan Goss Motor Cycles Limited of Yeovil who supplied the spare parts needed for the overhaul and to both the Avon Rubber Company and Amal Limited who kindly supplied some of the illustrations that accompany the text.

The cover photograph was arranged through the courtesy of Peter Chubb of Yeovil.

About this manual

The author of this manual has the conviction that the only way in which a meaningful and easy-to-follow text can be written is to carry out the work himself, under conditions similar to those found in the average household. As a result, the hands seen in the photographs are those of the author. Even the machines are not new: examples which have covered a considerable mileage are selected, so that the conditions encountered would be typical of those encountered by the average rider/owner. Unless specially mentioned, and therefore considered essential. Triumph service tools have not been used. There are invariably alternative means of slackening or removing some vital component when service tools are not available, but risk of damage is to be avoided at all costs.

Each of the eight chapters is divided into numbered sections. Within the sections are numbered paragraphs. Cross-reference throughout the manual is quite straightforward and logical. For example, when reference is made 'See Section 6.2' it means section 6, paragraph 2 in the same chapter. If another chapter were meant, the reference would read 'See Chapter 2, section 6.2'. All photographs are captioned with a section/paragraph number to which they refer, and are always relevant to the chapter text adjacent.

Figure numbers (usually line illustrations) appear in numerical order, within a given chapter. Fig. 1.1 therefore refers to the first figure in Chapter 1. Left hand and right hand descriptions of the machines and their component parts refer to the left and right when the rider is seated, facing forward.

Motorcycle manufacturers continually make changes to specifications and recommendations, and these, when notified, are incorporated into our manuals at the earliest opportunity.

Whilst every care is taken to ensure that the information in this manual is correct no liability can be accepted by the authors or publishers for loss, damage or injury caused by any errors in or omissions from the information given.

Contents

1960 350 cc Triumph Twenty-one (3TA)

1960 500 cc Triumph Speed Twin (5TA)

1969 500 cc Triumph Daytona (T100R)

Triumph Tiger 100 (T100C)

Introduction to the
Triumph 350/500cc unit-construction twins

Contrary to popular belief, the first Triumph vertical twin engine was designed and manufactured as far back as 1914, although a complete machine was not built. It was not until 1933 that another twin cylinder design emerged, this time in the form of a unit-construction 650 cc engine with geared primary drive, designed by Val Page. One of the completed models, with sidecar attached, covered 500 miles in 500 minutes at Brooklands immediately after the outfit had competed successfully in that year's International Six Days Trial. This feat, which was observed by the Auto-Cycle Union throughout, won the Maudes Trophy for Triumph Motors.

In January 1936 the Triumph Engineering Company Limited was formed to take over the motor cycle manufacturing activities of Triumph Motors. A new designer, Edward Turner, was appointed and it was Turner who inspired the Speed Twin model that had sensational impact on the motor cycle world during 1937. Indeed, this is the model that is the true ancestor of today's vertical twin designs and the one that established an entirely new trend in motor cycling.

After the war, the first models to re-appear were the Speed Twin and the Tiger 100, both virtually identical with the pre-war designs, apart from the use of telescopic forks and a separate dynamo and magneto in place of the Lucas Magdyno used previously. Soon, three new models were added to the range, a 350 cc version of the Speed Twin known as the model 3T, and two competition models, the Grand Prix racer and the Trophy Trials model. These latter models employed a variant of the generator engine developed by the Triumph Engineering Company during the war, which had an all-alloy cylinder barrel and head. Ernie Lyons had already won the first post-war Manx Grand Prix on a machine fitted with one of these engines and a prototype of the Triumph sprung hub. This latter device provided a means of giving the rear end of the rigid frame models a rudimentary form of suspension until the swinging arm layout went into production.

The manufacture of separate engine and gearbox models continued until 1957, when the first of the unit-construction models was announced at the end of that year. The first model to be marketed with the new unit-construction engine was a new 350 cc model, the 3TA or Twenty One, as it was named, to mark the twenty-first anniversary of the introduction of the Triumph vertical twin. Then followed a period of gradual modification until eventually all of the production models had unit-construction engines.

The Triumph engine has an outstanding reputation for reliability and high performance, two attributes that do not often go hand in hand. This is why the Triumph twin engine is invariably the one chosen by builders of 'specials', who wish to create their own machine by marrying together a collection of parts not necessarily from the one factory. Although the 350 cc and 500 cc models have now been phased out of production in favour of the larger capacity models, there is little doubt that they will continue to be popular for many years to come.

Guide to machine identification

Early models

Up to 1969 engine and frame numbers started with the 'H' prefix, the series ending at H67330. The model designation eg T100T, was added to indicate the machine's actual specification. Note that the engine and frame numbers should be the same.

Later models

From early 1969 the system of numbering was changed, and a prefix was added indicating the month and year of manufacture.

The first letter indicates the month of manufacture as follows:

A January   H July
B February   J August
C March   K September
D April   N October
E May   P November
G June   X December

The second letter indicates the year of manufacture as follows:

C 1969   G 1972
D 1970   H 1973
E 1971   J 1974

Note that some machines built in early 1969 were incorrectly stamped with the D year designation (AD07740 – 08884). Later models reverted to the C designation for that year.

The third Section is a numerical block of five figures which commenced 00100 in early 1969. These numbers started again at 00001 in 1971. The fourth Section indicates the model, eg T100R.

Note: Machines are identified at all times in this Manual by their Triumph year of production; this may not be the same as a machine's date of sale.

Ordering spare parts

When ordering spare parts for any of the Triumph unit-construction vertical twins, it is advisable to deal direct with an official Triumph agent who will be able to supply many of the items ex-stock. Parts cannot be obtained direct from the Triumph Engineering Company Limited; all orders must be routed through an approved agent, even if the parts required are not held in stock.

Always quote the engine and frame numbers in full. Include any letters before or after the number itself. The frame number will be found stamped on the left hand front down tube, adjacent to the steering head. The engine number is stamped on the left hand crankcase, immediately below the base of the cylinder barrel.

Use only parts of genuine Triumph manufacture. Pattern parts are available but in many instances they will have an adverse effect on performance and/or reliability. Some complete units are available on a 'service exchange' basis, affording an economic method of repair without having to wait for parts to be reconditioned. Details of the parts available, which include petrol tanks, front forks, front and rear frames, clutch plates, brake shoes etc. can be obtained from any Triumph agent. It follows that the parts to be exchanged must be acceptable before factory reconditioned replacements can be supplied.

Some of the more expendable parts such as spark plugs, bulbs, tyres, oils and greases etc., can be obtained from accessory shops and motor factors, who have convenient opening hours, charge lower prices and can often be found not far from home. It is also possible to obtain parts on a Mail Order basis from a number of specialists who advertise regularly in the motor cycle magazines.

MODEL:				T100R	T90	5TA	3TA
Basic dimensions:							
Wheel base	...	...	...	53½ in. (136 cm)	53½ in. (136 cm)	53½ in. (136 cm)	53½ in. (136 cm)
Overall length	...	...	...	83¼ in. (211.5 cm)	83¼ in. (211.5 cm)	83¼ in. (211.5 cm)	83¼ in. (211.5 cm)
Overall width	...	...	...	26½ in. (67.3 cm)	26½ in. (67.3 cm)	26½ in. (67.3 cm)	26½ in. (67.3 cm)
Overall height	...	...	...	38 in. (96.5 cm)	38 in. (96.5 cm)	38 in. (96.5 cm)	38 in. (96.5 cm)
Ground clearance	...	...	...	7½ in. (19 cm)	7½ in. (19 cm)	7½ in. (19 cm)	7½ in. (19 cm)
Weights:							
Unladen weight	...	...	...	337 lb (153 kgm)	336 lb (153 kgm)	340 lb (155 kgm)	340 lb (155 kgm)
Engine unit (dry)	...	...	...	106 lb (48 kgm)	104 lb (47.2 kgm)	106 lb (48 kgm)	104 lb (47.2 kgm)

Frame number location

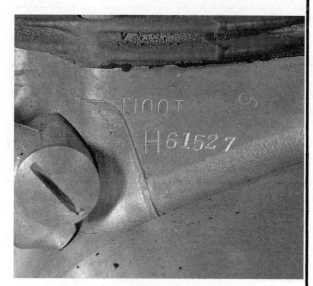

Engine number location

Routine maintenance

The need for weekly routine maintenance is something which cannot be over-emphasised. As soon as the machine is ridden every part is under some form of stress and it must be treated almost as though it were a living thing. If the machine is not used regularly, routine maintenance must still be carried out to prevent any components falling into decay.

Whilst keeping the machine in good condition, routine maintenance can prove invaluable as a kind of early warning system against failures of all kinds, mechanical and electrical. Charts and diagrams are shown to help and guide you, with dates and mileages. It should be remembered that the intervals between various maintenance tasks serve only as a guide. As the machine gets older or is perhaps subjected to harsh conditions, it is advisable to reduce the period between each check.

Some tasks are described in detail, but where they are not mentioned fully as routine maintenance items, they will be found elsewhere in the text under their appropriate chapters. No special tools are necessary for normal maintenance jobs, but those in the machine's tool kit coupled perhaps with those in the average garage at home should prove quite adequate for the task.

As well as keeping the bike in good condition mechanically, routine maintenance will pay dividends when you sell it. The Service Charts give no mention of brakes, which must be religiously checked and adjusted at all times. You are inclined to remember your brakes only when you need them, and if they have been neglected this may prove too late.

RM1 Engine oil must be changed every 3,000 miles

Weekly or every 250 miles (400 km)

Check level in oil tank and top up if necessary
Check level in primary chaincase
Lubricate rear chain
Check battery acid level and top up with distilled water if necessary
Check tyre pressures

Monthly or every 1000 miles (1600 km)

Change oil in primary chaincase
Lubricate all cables
Grease swinging arm fork pivot (on right hand end of spindle)
Remove, clean and lubricate final drive chain and re-adjust when refitted
Check and tighten nuts and bolts if necessary
Adjust tension of primary chain

Six weekly or every 1500 miles (2400 km)

Change oil in gearbox, also in front forks
Apply a light smear of grease to contact breaker cam
Check adjustment of steering head bearings

RM2 Battery is often the most neglected of all components

Three monthly or every 3000 miles (4800 km)

Change the engine oil and clean all filters in the lubrication system
Adjust points and clean and lubricate contact breaker heel and check ignition timing
Grease brake pedal spindle
Clean plugs and regap

Six monthly or every 6000 miles (9600 km)

Check oil level in gearbox
Examine front forks for oil leakage
Check valve clearances
Clean air filters and clean and adjust carburettor(s)

Yearly or every 12,000 miles (19,200 km)

Grease wheel bearings and steering head bearings

The 12,000 mile (19,200 km) service will mean a reasonable amount of dismantling (all details given in appropriate chapters). It should be noted that even when six-monthly and yearly maintenance has to be undertaken, the weekly, monthly, six-weekly and three-monthly services must be carried out in between. There is no part of a motor cycle's life when any routine maintenance tasks can be ignored or left out.

A few other items to be checked regularly:

The electrical system must be in good working order at all times
Tyres must be maintained at the correct pressure and adequately treaded for the job (ie solo or sidecar). They can be cut in a second so it is advisable to check regularly for signs of cracks or splits, bulging or uneven wear

All these things make common sense and are life-savers — someone else's and yours.

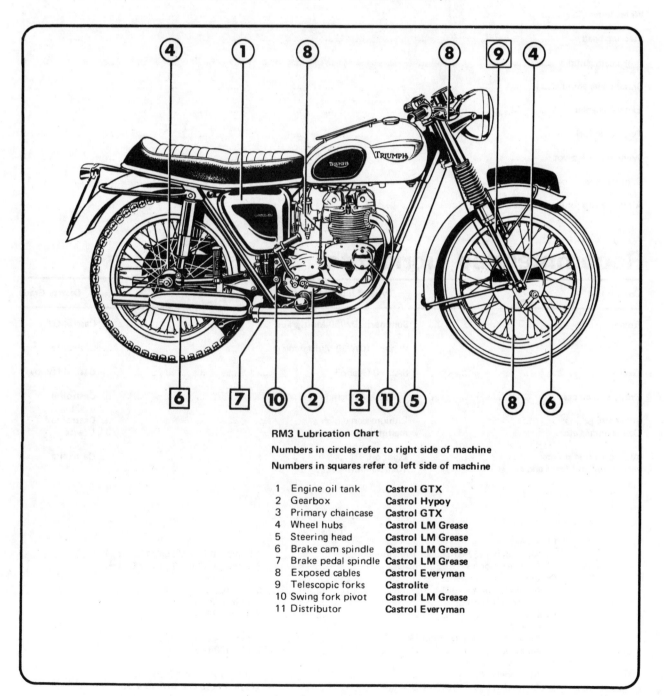

RM3 Lubrication Chart

Numbers in circles refer to right side of machine

Numbers in squares refer to left side of machine

1	Engine oil tank	Castrol GTX
2	Gearbox	Castrol Hypoy
3	Primary chaincase	Castrol GTX
4	Wheel hubs	Castrol LM Grease
5	Steering head	Castrol LM Grease
6	Brake cam spindle	Castrol LM Grease
7	Brake pedal spindle	Castrol LM Grease
8	Exposed cables	Castrol Everyman
9	Telescopic forks	Castrolite
10	Swing fork pivot	Castrol LM Grease
11	Distributor	Castrol Everyman

Guide to lubrication

Component	Key on illustration
Engine oil tank	1
Gearbox	2
Primary chaincase	3
Wheel hubs	4
Steering head	5
Brake cam spindle	6
Brake pedal spindle	7
Exposed cables	8
Telescopic fork	9
Swinging fork pivot	10
Contact breaker cam	11
All brake rod joints and pins	—

Recommended lubricants

Component	Type of lubricant	Castrol Grade
Engine	Summer: 20W/50 Multigrade	Castrol GTX
	Winter: 10W/30 Multigrade	Castrolite
Gearbox	SAE 90 Gear oil	Castrol Hypoy
Primary chaincase	10W/30 Multigrade	Castrolite
All grease points: Cam spindles etc.	Lithium based high melting point	Castrol LM Grease
All lubrication points: including front forks and cables	Light oil, 10W/30 Multigrade	Castrolite

Capacities

Fuel tank:
All models 1957-62 and all 1965 models ... ... 3.5 Imp gal (4.2 US gal, 16.0 litre)
All other models ... ... 3.0 Imp gal (3.6 US gal, 13.6 litre)
Oil tank:
All models up to 1965 ... ... 5.0 pints (2.84 litre)
All models 1966 on ... ... 5.5 pints (3.13 litre)
Gearbox ... ... 2/3 pint (379 cc)
Primary chaincase:
All models after KD27850 (Sept 1970) ... ... ¼ pint (142 cc) - initial fill only
All earlier models ... ... ½ pint (284 cc)
Front forks - per leg:
All models up to 1963 ... ... ¼ pint (142 cc) SAE 20 or 30
All models after H32465 (1964 on) ... ... 1/3 pint (190 cc) SAE 10W/30

List of illustrations

Safety first!

Professional motor mechanics are trained in safe working procedures. However enthusiastic you may be about getting on with the job in hand, do take the time to ensure that your safety is not put at risk. A moment's lack of attention can result in an accident, as can failure to observe certain elementary precautions.

There will always be new ways of having accidents, and the following points do not pretend to be a comprehensive list of all dangers; they are intended rather to make you aware of the risks and to encourage a safety-conscious approach to all work you carry out on your vehicle.

Essential DOs and DON'Ts

DON'T start the engine without first ascertaining that the transmission is in neutral.

DON'T suddenly remove the filler cap from a hot cooling system – cover it with a cloth and release the pressure gradually first, or you may get scalded by escaping coolant.

DON'T attempt to drain oil until you are sure it has cooled sufficiently to avoid scalding you.

DON'T grasp any part of the engine, exhaust or silencer without first ascertaining that it is sufficiently cool to avoid burning you.

DON'T allow brake fluid or antifreeze to contact the machine's paintwork or plastic components.

DON'T syphon toxic liquids such as fuel, brake fluid or antifreeze by mouth, or allow them to remain on your skin.

DON'T inhale dust – it may be injurious to health (see *Asbestos* heading).

DON'T allow any spilt oil or grease to remain on the floor – wipe it up straight away, before someone slips on it.

DON'T use ill-fitting spanners or other tools which may slip and cause injury.

DON'T attempt to lift a heavy component which may be beyond your capability – get assistance.

DON'T rush to finish a job, or take unverified short cuts.

DON'T allow children or animals in or around an unattended vehicle.

DON'T inflate a tyre to a pressure above the recommended maximum. Apart from overstressing the carcase and wheel rim, in extreme cases the tyre may blow off forcibly.

DO ensure that the machine is supported securely at all times. This is especially important when the machine is blocked up to aid wheel or fork removal.

DO take care when attempting to slacken a stubborn nut or bolt. It is generally better to pull on a spanner, rather than push, so that if slippage occurs you fall away from the machine rather than on to it.

DO wear eye protection when using power tools such as drill, sander, bench grinder etc.

DO use a barrier cream on your hands prior to undertaking dirty jobs – it will protect your skin from infection as well as making the dirt easier to remove afterwards; but make sure your hands aren't left slippery. Note that long-term contact with used engine oil can be a health hazard.

DO keep loose clothing (cuffs, tie etc) and long hair well out of the way of moving mechanical parts.

DO remove rings, wristwatch etc, before working on the vehicle – especially the electrical system.

DO keep your work area tidy – it is only too easy to fall over articles left lying around.

DO exercise caution when compressing springs for removal or installation. Ensure that the tension is applied and released in a controlled manner, using suitable tools which preclude the possibility of the spring escaping violently.

DO ensure that any lifting tackle used has a safe working load rating adequate for the job.

DO get someone to check periodically that all is well, when working alone on the vehicle.

DO carry out work in a logical sequence and check that everything is correctly assembled and tightened afterwards.

DO remember that your vehicle's safety affects that of yourself and others. If in doubt on any point, get specialist advice.

IF, in spite of following these precautions, you are unfortunate enough to injure yourself, seek medical attention as soon as possible.

Asbestos

Certain friction, insulating, sealing, and other products – such as brake linings, clutch linings, gaskets, etc – contain asbestos. *Extreme care must be taken to avoid inhalation of dust from such products since it is hazardous to health.* If in doubt, assume that they *do* contain asbestos.

Fire

Remember at all times that petrol (gasoline) is highly flammable. Never smoke, or have any kind of naked flame around, when working on the vehicle. But the risk does not end there – a spark caused by an electrical short-circuit, by two metal surfaces contacting each other, by careless use of tools, or even by static electricity built up in your body under certain conditions, can ignite petrol vapour, which in a confined space is highly explosive.

Always disconnect the battery earth (ground) terminal before working on any part of the fuel or electrical system, and never risk spilling fuel on to a hot engine or exhaust.

It is recommended that a fire extinguisher of a type suitable for fuel and electrical fires is kept handy in the garage or workplace at all times. Never try to extinguish a fuel or electrical fire with water.

Note: *Any reference to a 'torch' appearing in this manual should always be taken to mean a hand-held battery-operated electric lamp or flashlight. It does **not** mean a welding/gas torch or blowlamp.*

Fumes

Certain fumes are highly toxic and can quickly cause unconsciousness and even death if inhaled to any extent. Petrol (gasoline) vapour comes into this category, as do the vapours from certain solvents such as trichloroethylene. Any draining or pouring of such volatile fluids should be done in a well ventilated area.

When using cleaning fluids and solvents, read the instructions carefully. Never use materials from unmarked containers – they may give off poisonous vapours.

Never run the engine of a motor vehicle in an enclosed space such as a garage. Exhaust fumes contain carbon monoxide which is extremely poisonous; if you need to run the engine, always do so in the open air or at least have the rear of the vehicle outside the workplace.

The battery

Never cause a spark, or allow a naked light, near the vehicle's battery. It will normally be giving off a certain amount of hydrogen gas, which is highly explosive.

Always disconnect the battery earth (ground) terminal before working on the fuel or electrical systems.

If possible, loosen the filler plugs or cover when charging the battery from an external source. Do not charge at an excessive rate or the battery may burst.

Take care when topping up and when carrying the battery. The acid electrolyte, even when diluted, is very corrosive and should not be allowed to contact the eyes or skin.

If you ever need to prepare electrolyte yourself, always add the acid slowly to the water, and never the other way round. Protect against splashes by wearing rubber gloves and goggles.

Mains electricity

When using an electric power tool, inspection light etc which works from the mains, always ensure that the appliance is correctly connected to its plug and that, where necessary, it is properly earthed (grounded). Do not use such appliances in damp conditions and, again, beware of creating a spark or applying excessive heat in the vicinity of fuel or fuel vapour.

Ignition HT voltage

A severe electric shock can result from touching certain parts of the ignition system, such as the HT leads, when the engine is running or being cranked, particularly if components are damp or the insulation is defective. Where an electronic ignition system is fitted, the HT voltage is much higher and could prove fatal.

Chapter 1 Engine

Contents

Specifications

Model T100R - Daytona and T100C - Trophy 500

Basic details

Bore and stroke	69 x 65.5 mm
Bore and stroke	2.7165 x 2.58 in
Cubic capacity	490 cc (30 cu in)
Compression ratio	9.0 : 1
Capacity of combustion chamber	31 cc

Crankshaft

Type	Forged two throw crank with bolt on flywheel
Left main bearing - size and type	72 x 30 x 90 mm Roller bearing
Left crankshaft main bearing journal diameter	1.1805 - 1.1808 in (29.985 - 29.992 mm)
Left bearing housing diameter...	2.8336 - 2.8321 in (71.973 - 71.935 mm)
Right main bearing - size and type	72 x 35 x 17 mm Ball bearing
Right crankshaft main bearing journal diameter	1.3774 - 1.3777 in (34.986 - 34.994 mm)
Right bearing housing diameter	2.8336 - 2.8321 in (71.973 - 71.935 mm)
Big end journal diameter	1.4375 - 1.4380 in (36.513 - 36.525 mm)
Minimum regrind diameter	1.4075 - 1.4080 in (35.751 - 35.763 mm)
Crankshaft end float	0.008 - 0.017 in (0.2032 - 0.4318 mm)

Connecting rods
Material ... ...   Alloy 'H' section RR 56
Length (centres) ... ...   5.311 - 5.313 in (134.899 - 134.95 mm)
Big end bearing type ... ...   Steel backed white metal
Bearing side clearance ... ...   0.013 - 0.017 in (0.3302 - 0.4018 mm)
Bearing diametral clearance ... ...   0.0005 - 0.0020 in min (0.0127 - 0.0508 mm)

Gudgeon pin
Material ... ...   High tensile steel
Diameter ... ...   0.6882 - 0.6885 in (17.48 - 17.49 mm)
Length ... ...   2.151 - 2.156 in (54.635 - 54.76 mm)

Small end (no bush)
Diameter ... ...   0.689 - 0.6894 in (17.5 - 17.508 mm)

Pistons
Material ... ...   Aluminium alloy die casting

	From H 49833	Before H 49833
Clearance:		
Top of skirt	0.0030 - 0.0045 in (0.1270 - 0.183 mm)	0.0075 - 0.0085 in (0.1905 - 0.2159 mm)
Bottom of skirt	0.0030 - 0.0045 in (0.0762 - 0.1143 mm)	0.002 - 0.003 in (0.0508 - 0.0762 mm)
Gudgeon pin hole diameter	0.6882 - 0.6886 in (17.48 - 17.49 mm)	0.6882 - 0.6886 in (17.48 - 17.49 mm)

Piston rings
Material ... ...   Cast iron
Compression rings (taper faced):
  Width ... ...   0.0615 - 0.0625 in (1.562 - 1.587 mm)
  Thickness ... ...   0.092 - 0.100 in (2.34 - 2.54 mm)
  Fitted gap ... ...   0.010 - 0.014 in (0.254 - 0.356 mm)
  Clearance in groove ... ...   0.001 - 0.003 in (0.0254 - 0.076 mm)
Oil control ring:
  Width ... ...   0.124 - 0.125 in (3.15 - 3.18 mm)
  Thickness ... ...   0.092 - 0.100 in (2.34 - 2.54 mm)
  Fitted gap ... ...   0.010 - 0.014 in (0.254 - 0.356 mm)
  Clearance in groove ... ...   0.0005 - 0.0025 in (0.01270 - 0.0635 mm)

Valves
Seat angle (included) ... ...   90°
Head diameter:
  Inlet ... ...   1 17/32 in (38.89 mm)
  Inlet (before H 49833) ... ...   1 7/16 in (36.5 mm)
  Exhaust ... ...   1 5/16 in (33.34 mm)
Stem diameter:
  Inlet ... ...   0.3095 - 0.3100 in (7.86 - 7.87 mm)
  Exhaust ... ...   0.3090 - 0.3095 in (7.849 - 7.86 mm)

Valve guides
Material ... ...   Hidural
Bore diameter (inlet and exhaust) ... ...   0.312 - 0.313 in (7.925 - 7.95 mm)
Outside diameter (inlet and exhaust) ... ...   0.5005 - 0.5010 in (12.713 - 12.75 mm)
Length - inlet and exhaust ... ...   1.760 - 1.770 in (44.85 - 44.96 mm)

Valve springs (Inner — Yellow, Outer — L/Blue Spot)

	Outer	Inner
Free length	1½ in (38.1 mm)	1 19/32 in (40.48 mm)
Total number of coils	6 (152.4)	8¼ (209.55)
Total fitted load:		
Valve open	136 lb (60.8 kg)	
Valve closed	63 lb (28.1 kg)	

Valve timing
Inlet opens ... ...   40° BTDC
Inlet closes ... ...   52° ABDC   Set all tappet clearances at
Exhaust opens ... ...   61° BBDC   0.020 in (0.5 mm) for checking
Exhaust closes ... ...   31° ATDC

Rockers
Material ... ...   High tensile steel forging
Bore diameter ... ...   0.4375 - 0.4380 in (11.113 - 11.125 mm)

Rocker spindle diameter ... ...   0.4355 - 0.4360 in (11.06 - 11.07 mm)
Tappet clearance (cold):
   Inlet ... ...   0.002 in (0.05 mm)
   Exhaust ... ...   0.004 in (0.10 mm)

Tappets
   Material ... ...   High tensile steel forging — Stellite Tip
   Tip radius ... ... $1\frac{1}{8}$ in (28.57 mm)
   Tappet diameter ... ...   0.3110 - 0.3115 in (7.89 - 7.912 mm)
   Clearance in guide block ... ...   0.0005 - 0.0015 in (0.0127 - 0.038 mm)

Tappet guide block
   Diameter of bores ... ...   0.3120 - 0.3125 in (7.925 - 7.938 mm)
   Outside diameter ... ...   1.000 - 0.9995 in (25.4 - 25.27 mm)
   Interference fit in cylinder block ... ...   0.0005 - 0.0015 in (0.0127 - 0.038 mm)

Camshafts
   Journal diameter - left ... ...   0.8100 - 0.8105 in (20.6 - 20.58 mm)
   Diametral clearance - left ... ...   0.0010 - 0.0025 in (0.0254 - 0.0635 mm)
   End float ... ...   0.005 - 0.008 in (0.127 - 0.2032 mm)
   Cam lift:
      Inlet ... ...   0.314 in (7.98 mm)
      Exhaust ... ...   0.314 in (7.98 mm)
   Base circle diameter - inlet and exhaust ... ...   0.812 in (20.62 mm)

Camshaft bearing bushes
   Material ... ...   Steel backed bronze
   Bore diameter (fitted) - left ... ...   0.8125 - 0.8135 in (20.64 - 20.663 mm)
   Outside diameter - left ... ...   0.906 - 0.907 in (23.01 - 23.05 mm)
   Length:
      Left inlet ... ...   1.114 - 1.094 in (28.29 - 27.78 mm)
      Left exhaust ... ...   0.922 - 0.942 in (23.42 - 23.93 mm)
   Interference fit in crankcase - left ... ...   0.002 - 0.003 in (0.05 - 0.076 mm)

Timing gears
   Inlet and exhaust camshaft pinions:
      No of teeth   50
      Interference fit on camshaft ... ...   0.000 - 0.001 in (0.000 - 0.0254 mm)
   Intermediate timing gear:
      No of teeth   42
      Bore diameter ... ...   0.5618 - 0.5625 in (14.27 - 14.29 mm)
   Intermediate timing gear bush:
      Material ... ...   Phosphor bronze
      Outside diameter ... ...   0.5635 - 0.5640 in (14.31 - 14.33 mm)
      Bore diameter ... ...   0.4990 - 0.4995 in (12.685 - 12.687 mm)
      Length ... ...   0.6775 - 0.6825 in (17.31 - 17.34 mm)
      Working clearance on spindle ... ...   0.0005 - 0.0015 in (0.0127 - 0.038 mm)
   Intermediate wheel spindle:
      Diameter ... ...   0.4980 - 0.4985 in (12.65 - 12.66 mm)
      Interference fit in crankcase ... ...   0.0005 - 0.0015 in (0.0127 - 0.038 mm)
   Crankcase pinion:
      No of teeth   25
      Fit on crankcase ... ...   + 0.0003 in (0.00762 mm) — 0.0005 in (0.0127 mm)

Cylinder block
   Material ... ...   Cast iron
   Bore size ... ...   2.7160 - 2.7165 in (68.98 - 68.99 mm)
   Maximum oversize ... ...   2.7360 - 2.7365 in (69.49 - 69.5 mm)
   Tappet guide block housing diameter ... ...   0.9985 - 0.9990 in (25.36 - 25.37 mm)

Cylinder head
   Material ... ...   DTD 424 Aluminium alloy
   Inlet port size ... ... $1\frac{1}{16}$ in dia (26.99 mm)
   Exhaust port size ... ...   1¼ in dia (31.75 mm)
   Valve seatings:
      Type ... ...   Cast-in
      Material ... ...   Cast iron

Model T100S — Tiger 100

Crankshaft
   Left main bearing, size and type ... ...   72 x 30 x 19 mm Ball journal
   Right crankshaft main bearing journal diameter ... ...   1.4375 - 1.4380 in (36.513 - 36.525 mm)

Right main bearing bore, size and type ... ...   1.4385 - 1.4390 in (36.538 - 36.551 mm) Steel backed copper lead
                                                                   lined bush. Under sizes available: —0.010 in, —0.020 in, —0.030 in

Right main bearing housing diameter   ... ...   1.8135 - 1.8140 in (46.06 - 46.07 mm)

Gudgeon pin
Fit in small end bush   ... ...   0.0005 - 0.0012 in (0.0127 - 0.0305 mm)

Small end bush
Material   ... ...   Phosphor bronze
Outer diameter   ... ...   0.782 - 0.783 in (19.863 - 19.888 mm)
Length   ... ...   0.890 - 0.910 in (22.61 - 23.1 mm)
Finished bore diameter ... ...   0.6905 - 0.6910 in (17.54 - 17.55 mm)

Valve timing
Inlet opens   ... ...   34⁰ BTDC
Inlet closes   ... ...   55⁰ ABDC   Set all tappet clearances at
Exhaust opens   ... ...   48⁰ BBDC   0.020 in (0.5 mm) for checking
Exhaust closes   ... ...   27⁰ ATDC

Tappets
Tip radius   ... ...   ¾ in (19.05 mm)

Camshafts
Cam lift (exhaust)   ... ...   0.296 in (7.518 mm)

Cylinder head
Inlet port size   ... ...   1 in (25.4 mm)

Model T100T

Basic details
Compression ratio   ... ...   9 : 1
Power output (bhp at rpm)   39 at 7400

Valves
Head diameter inlet ... ...   1 17/32 in (38.89 mm)
Head diameter exhaust ... ...   1 5/16 in (33.33 mm)

Pistons
Clearance:
Top of skirt   ... ...   0.0050 - 0.0072 in (0.127 - 0.183 mm)
Bottom of skirt   ... ...   0.0030 - 0.0045 in (0.076 - 0.114 mm)

Valve timing
Inlet opens   ... ...   40⁰ BTDC
Inlet closes   ... ...   52⁰ ABDC   Set all tappet clearances at
Exhaust opens   ... ...   61⁰ BBDC   0.020 in (0.5 mm) for checking
Exhaust closes   ... ...   31⁰ ATDC

Tappets
Tip radius   ... ...   1 1/8 in (28.57 mm)

Camshafts
Cam lift:
Inlet   ... ...   0.314 in (8.98 mm)
Exhaust   ... ...   0.314 in (8.98 mm)

Timing gears
Inlet and exhaust camwheels ... ...   3 keyway

Cylinder head
Inlet port size   ... ...   1 1/16 in (26.98 mm)

Model 5TA – Speed Twin (discontinued after engine number H 49833)

Basic details
Bore and stroke   ... ...   69 x 65.5 mm
Bore and stroke   ... ...   2.7165 x 2.58 in
Cubic capacity   ... ...   490 cc (30 cu in)
Compression ratio   ... ...   7 : 1
Capacity of combustion chamber   ... ...   35 cc (2.14 cu in)
Power output (bhp at rpm)   ... ...   27 at 6500

Pistons

Material	Aluminium alloy die casting
Clearance:	
Top of skirt	0.0065 - 0.0075 in (0.165 - 0.190 mm)
Bottom of skirt	0.001 - 0.002 in (0.0254 - 0.0508 mm)
Gudgeon pin hole diameter	0.6882 - 0.6886 in (17.45 - 17.46 mm)

Valve timing

Set all tappet clearances at 0.020 in (0.5 mm) for checking

	1963 on	**Pre 1963**
Inlet opens	34° BTDC	26.5° BTDC
Inlet closes	55° ABDC	69.5° ABDC
Exhaust opens	48° BBDC	61.5° BBDC
Exhaust closes	27° ATDC	35.5° ATDC

Valves

Seat angle (included)	90°
Head diameter:	
Inlet	1 7/16 in (36.51 mm)
Exhaust	1 5/16 in (33.34 mm)

Valve guides

Material	Cast iron
Bore diameter (inlet and exhaust)	0.3130 - 0.3120 in (7.91 - 7.88 mm)
Outside diameter (inlet and exhaust)	0.5005 - 0.5010 in (12.71 - 12.73 mm)
Length (inlet and exhaust)	1 3/4 in (44.45 mm)

Model T90 - Tiger 90

Basic details

Bore and stroke	2.2928 x 2.58 in (58.25 x 65.5 mm)
Cubic capacity	349 cc (21 cu in)
Compression ratio	9.5 : 1
Capacity of combustion chamber	19.5 cc (1.19 cu in)
Power output (bhp - rpm)	27 at 7500

Pistons

Material	Aluminium alloy die casting	
	From H 49833	**Before H 49833**
Clearance:		
Top of skirt	0.0049 - 0.0070 in (0.114 - 0.178 mm)	0.0058 - 0.0068 in (0.1473 - 0.173 mm)
Bottom of skirt	0.0034 - 0.0049 in (0.0864 - 0.114 mm)	0.0043 - 0.0048 in (0.109 - 0.121 mm)
Gudgeon pin hole diameter	0.5618 - 0.5621 in (14.269 - 14.277 mm)	0.5618 - 0.5621 in (14.269 - 14.277 mm)

Valves

Seat angle (inlet and exhaust)	90° included angle
Head diameter:	
Inlet	1 7/16 in (36.5 mm)
Exhaust	1 3/16 in (30.1 mm)

Valve guides

Material	Hidural
Length (inlet and exhaust)	1 3/4 in (44.45 mm)

Valve timing

Set all tappet clearances at 0.020 in (0.5 mm) for checking

Inlet opens	34° BTDC
Inlet closes	55° ABDC
Exhaust opens	48° BBDC
Exhaust closes	27° ATDC

Cylinder head

Material	DTD 424 Aluminium alloy
Inlet port size	1 in dia (25.4 mm)
Exhaust port size	1 1/4 in dia (31.75 mm)
Valve seatings:	
Type	Cast-in
Material	Cast iron

Connecting rods

Material	'H' section RR 56 alloy
Material (before H 49833)	EN 18 steel stamping

3TA — Twenty-One (discontinued after engine number H 49833)

Basic details

Bore and stroke	58.25 x 65.5 mm
Bore and stroke	2.2928 x 2.58 in
Cubic capacity	349 cc (21 cu in)
Compression ratio	7.5 : 1
Capacity of combustion chamber	23.4 cc (1.42 cu in)
Power output (bhp at rpm)	18.5 at 6500

Cylinder head

Material	DTD 424 Aluminium alloy
Inlet port size	7/8 in (22.23 mm)
Exhaust port size	1 3/16 in (30.2 mm)
Valve seatings:	
Type	Cast-in
Material	Cast iron

Pistons

Material	Aluminium alloy die casting
Clearance:	
Top of skirt	0.0048 - 0.0058 in (0.122 - 0.147 mm)
Bottom of skirt	0.0033 - 0.0043 in (0.084 - 0.109 mm)
Gudgeon pin hole diameter	0.5618 - 0.5621 in (14.27 - 14.28 mm)

Valves

Seat angle (inlet and exhaust)	90° included angle
Head diameter:	
Inlet	1 5/16 in dia (33.34 mm)
Exhaust	1 3/16 in dia (30.2 mm)
Stem diameter:	
Inlet	0.3095 - 0.3100 in (7.86 - 7.87 mm)
Exhaust	0.3090 - 0.3095 in (7.85 - 7.86 mm)

Valve guides

Material	Cast iron
Bore diameter (inlet and exhaust)	0.3120 - 0.3130 in (8.925 - 8.95 mm)
Outside diameter (inlet and exhaust)	0.5005 - 0.5010 in (12.713 - 12.725 mm)
Length:	
Inlet and exhaust	1¾ in (94.45 mm)

Valve timing

Inlet opens	26½° BTDC	
Inlet closes	69½° ABDC	Set all tappet clearances at
Exhaust opens	61½° BBDC	0.020 in (0.5 mm) for checking
Exhaust closes	35½° ATDC	

Rockers

Material	High tensile steel forging
Bore diameter	0.4375 - 0.4390 in (11.13 - 11.151 mm)
Rocker spindle diameter	0.4355 - 0.4360 in (10.062 - 11.07 mm)
Tappet clearance (cold):	
Inlet and exhaust	0.010 in (0.25 mm)

Camshafts

Journal diameter - left	0.8100 - 0.8105 in (20.57 - 20.58 mm)
Diametral clearance - left	0.0010 - 0.0025 in (0.0254 - 0.0635 mm)
End float	0.005 - 0.008 in (0.127 - 0.2032 mm)
Cam lift:	
Inlet	0.281 in (7.14 mm)
Exhaust	0.281 in (7.14 mm)
Base circle diameter	0.812 in (20.62 mm)

Unless otherwise stated, T100R data applies to all models

Torque wrench settings (dry)

	lb f ft	kg f m
Flywheel bolts	33	4.24
Connecting rod bolts	18	2.489
Crankcase junction bolts	15	2.074
Crankcase junction studs	20	2.765
Cylinder block nuts	35	4.835
Cylinder head bolts (3/8 in diameter)	18	2.489
Rocker box nuts	5	0.69
Rocker box bolts	5	0.69
Rocker spindle domed nuts	25	3.318
Oil pump nuts	6	0.83
Kickstart ratchet pinion nut	40	5.53
Clutch centre nut	50	6.9
Rotor fixing nut	30	4.148
Stator fixing nuts	20	2.765
Headlamp pivot bolts	10	1.38
Headrace sleeve nut pinch bolt	15	2.074
Stanchion pinch bolts	25	3.318
Front wheel spindle cap bolts	25	3.318
Brake cam spindle nuts	20	2.765
Zener diode fixing nut	1½	0.207
Twin carburettor manifold socket screws	10	1.38
Fork restrictor bolt	8	1.106

1 General description

The engine fitted to the Triumph unit-construction vertical twins is of the combined engine and gearbox type, in which the gearbox casting forms an integral part of the right hand crankcase and the primary chaincase, an integral part of the left hand chaincase. Aluminium alloy is used for all of the engine castings, with the exception of the cylinder barrel, which is of cast iron.

The cylinder head has cast-in austenitic valve seat inserts and houses the overhead valves which are actuated by rocker arms enclosed within the detachable rocker boxes. The pushrods are of aluminium alloy with hardened end pieces.

'H' section connecting rods of hinduminium alloy, with detachable caps and steel-backed shell bearings, carry aluminium alloy die-cast pistons, each with two compression rings and one oil scraper ring. The two-throw crankshaft has a detachable shrunk-on cast iron flywheel, retained in a central position by three high tensile steel bolts. The cast iron cylinder barrel houses the press-fit tappet guide blocks.

The separate inlet and exhaust camshafts operate in sintered bronze bushes, mounted transversely in the upper part of the crankcase. The camshafts are driven by the train of timing gears, from the right hand end of the crankshaft. The inlet camshaft provides the drive for the oil pump and the rotary breather valve disc, whilst the exhaust camshaft drives the contact breaker and on some models, the tachometer drive gearbox.

Power from the engine is transmitted in the conventional manner through the engine sprocket and primary chain to the clutch unit that incorporates a shock absorber. Chain tension is controlled by an adjustable chain tensioner, immersed within the oil content of the chaincase.

2 Operations with engine in frame

The Daytona and 500 engines can be worked on quite easily in the frame for a number of operations, excluding the obvious, like main bearings, crankshaft etc. Operations that can be carried out with the engine in the frame are as follows:

1 Removing and replacing the cylinder head.
2 Removing and replacing the barrels and pistons.
3 Removing and replacing the alternator.
4 Removing and replacing the primary drive components.
5 Removing and replacing the oil pump and points.

3 Method of engine and gearbox removal

If a major rebuild is necessary, the engine unit will have to be removed to facilitate splitting the crankcase for access to the big ends, main bearings, crankshaft etc. This is a one-man job until lifting out the engine/gearbox assembly is necessary. At a weight of 106 lbs (48 kg) it is fairly heavy, so some assistance is advised at this point in the proceedings.

4 Removing the engine/gearbox unit

1 Before attempting to take the engine/gearbox unit out of the frame it is recommended that the machine should be cleaned with Gunk or similar cleaning agent.
2 Place the machine on the centre stand on firm, level ground. If it is at all unsteady, use wood to stabilise it.
3 Drain the oil from the oil tank, using a catchment tank of at least 5½ pints capacity. Remove the drain plug or on earlier models, the main feed pipe and leave it to drain whilst the rest of the stripdown is undertaken.
4 Turn the petrol taps to 'off' position and undo the hexagon nuts adjoining the taps. Undo the two front nuts under the tank (some American export models have reflectors fitted that are also retained by these nuts). Undo the single bolt at the rear of the tank passing through a welded lug. Remove the tank, taking care not to lose any rubbers from the mountings.
5 Remove the battery connections and the battery, and place it in a safe place (if removed for some time a periodical trickle charge (1 amp) should be given to keep the battery in good condition).
6 Undo the clamp bolt on the kickstarter; the gear lever bolt will have to be extracted to permit removal, then replace the bolt in the freed levers. Place all parts in a box, a few of which should be available when stripping the machine.

7   Unfasten the bolts on the brackets of the front inner exhaust pipes. Undo the clamp bolts on the finned clamps adjoining the head, also unfasten the pillion footrests which hold the silencers to the machine. Remove the exhaust system - on some models a balance pipe is used; this is clamped on and the bolts must be thoroughly slackened to enable the exhaust system to be removed.

8   Unscrew and remove the tachometer cable located in front of the cylinder barrels on the crankcase.

9   Unfasten the screws which hold the clips which clamp the air filters (some models have screw-on fittings) and remove.

10   Remove the screw on the rear brake rod to disconnect the brake light spring. Undo the nut on the end of the footbrake spindle.

11   Unfasten the bolts holding the footrests and remove them both.

12   Slacken off the clutch adjustment at the handlebar lever and withdraw the rubber cover from the clutch abutment on the gearbox. Unscrew the abutment and detach the cable.

13   Pull off the plug leads and tape them out of the way. It is also recommended that the coils be unclamped and placed out of the way, taped to the top of the frame.

14   Using a Phillip's screwdriver, remove the four screws holding on the carburettor tops (two on each carburettor). It is recommended that a rag should be wrapped around the slides to protect their working surfaces. Tape them out of the way.

15   Disconnect the two wires which connect the contact breaker points to the wiring loom (they are colour coded for easy recognition).

16   Disconnect the snap connectors under the gearbox from their connection with the alternator.

17   Take off the small bridging chainguard cover and disconnect the spring link of the chain; pull off the chain (the machine must be in neutral).

18   Take off the nuts which hold the head steady, making sure that the spacers are not dropped or lost. The widest spacer is usually on the left side.

19   Unfasten the domed nuts holding on the oil feed pipes. Take great care not to bend the pipes as they are liable to fracture. Tape them out of the way.

20   Withdraw the front engine plate bolts and remove the engine plates.

21   Slacken the long bolt which passes through the frame and the underside of the engine. Do not withdraw it yet.

22   Remove the rear engine plates situated above the gearbox.

23   After removal, all the engine bolts should have their nuts spun on lightly and arranged in order to aid reassembly.

24   Clear a good space for the engine, and if a complete stripdown is to follow, it is suggested that paper or rag be spread on the bench.

25   The engine is now retained by the long bolt which passes through the bottom of the crankcase. With one person supporting the engine and another taking the weight of the unit, extract the long bolt, taking care not to lose the spacers. Note their location for reassembly. Lift the engine out from the left hand side of the frame.

5   Dismantling the engine - general

1   The engine should be cleaned, preferably before commencing the stripdown. Care must be taken not to allow any cleaning agent into the exhaust inlets or any exposed parts.

2   Never use undue force to remove any stubborn part. Unless mention is made of the fact that there may be any special difficulty, it may be assumed that the dismantling operation had been tackled in the wrong sequence.

3   Dismantling is easier if an engine stand is constructed. This will allow both hands to be free, with the engine held steady on the bench.

6   Dismantling the engine - removing the cylinder head, barrel and pistons

1   Remove the carburettors by removing the flange nuts and bolts which hold the carburettors to the manifolds.

2   Remove the rocker caps.

3   Unscrew the nuts on the underside of the rocker boxes which will also release the rocker cap retaining springs.

4   Unscrew the Phillips screws - two of which are located on top of each rocker box.

5   Unfasten the two long bolts. The rocker boxes will rise slightly at this point because one valve is open.

6   Both rocker boxes can now be removed. Lift out the pushrods.

7   The cylinder head is removed by withdrawing the four remaining bolts on top of the head. The cylinder head gasket can be re-used if it is not damaged and if it is of the solid copper type. Pull off the pushrod tubes.

4.3 Drain oil from pipes

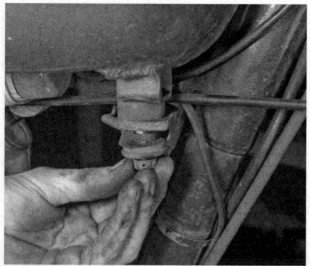

4.4 Front mounting bolts are wired together

4.4a Rear bolt passes through tank lug

4.7 Stays support front of exhaust pipes

4.7a Silencer is retained by pillion footrest

4.14 Protect slide surfaces, after removal

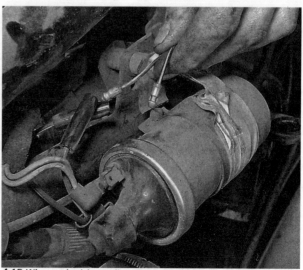

4.15 Wires to ignition coils are colour-coded

4.17 Small bridge chainguard protects chain at joint

4.18 Note position of stays when removing head steady

4.19 Check pipes do not twist when slackening union nuts

4.20 Remove front engine plates completely

4.25 Note location of spacer on lower engine bolt

4.25a Lift out engine unit from left-hand side

8 The cylinder barrel is removed by unfastening the nuts around the base. Be very careful not to lose the washers under the nuts. A thin spanner is needed here - there is usually one provided in the tool kit for the motor cycle.

9 Before lifting off the cylinder barrel, either elastic bands or rubber blocks should be wrapped around or jammed in the cam followers to stop them from dropping out when the barrel is lifted.

10 Turn the engine over until the pistons are at top dead centre (ie the pistons are at the top of the bores). Pull the cylinder barrel up until just clear of the crankcase and place rag around the crankcase mouth to prevent any bits of piston ring, which could be broken, from dropping into the crankcase. Pull off the barrels slowly, firmly and squarely.

11 Polythene tube cut into approximately 1 inch lengths should be pushed over the cylinder barrel studs to stop the piston working surfaces being damaged, if they come into contact with them.

12 Before placing the cylinder barrel to one side, take out the cam followers and mark them. They have to be put back in identical order otherwise rapid wear will occur and create a very noisy engine.

13 Remove the wire circlip from each piston and push out each gudgeon pin. If the pin is tight you may need to use an alloy drift and tap out the pin, taking great care not to damage the piston or connecting rod. Triumph service tool Z72 is available to extract them. Always support the piston when any pressure is exerted on either the connecting rod or gudgeon pin.

14 Discard the circlips as they should never be re-used. New replacements are essential to eliminate the possibility of them working loose whilst the engine is running, and so causing the gudgeon pin to hit the cylinder bore to create a great deal of damage.

15 Mark each piston on the inside to ensure that it is replaced in the same bore from which it came, facing the same direction. Failing to refit them correctly will result in loss of engine power and very high engine oil consumption.

6.1 Note balance pipe linking carburettor manifolds

6.3 Spring clip is released when nut is removed

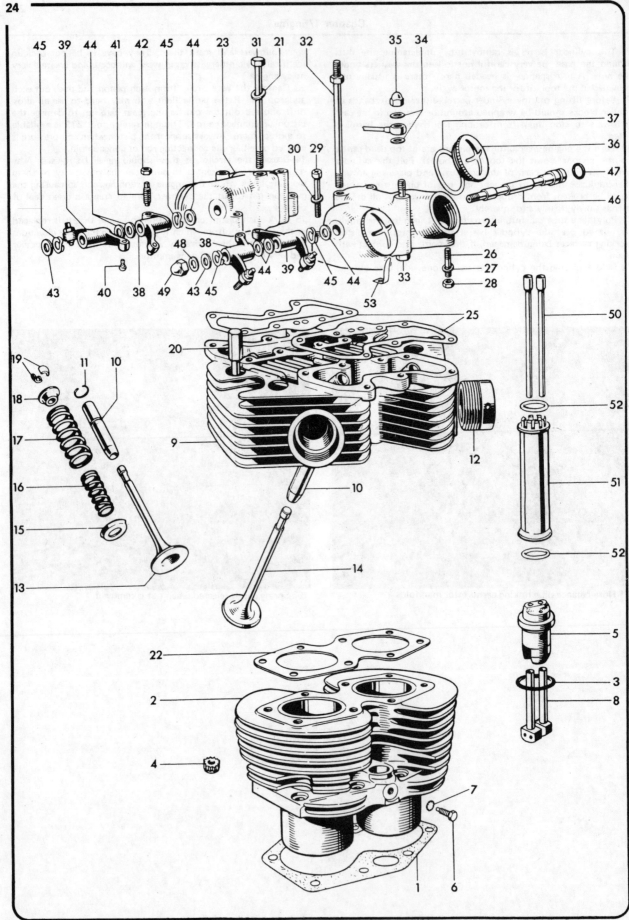

Fig. 1.1. Cylinder barrel and head, T100R and T100T models

1	Cylinder base washer 1 off		28	Nut 4 off
2	Cylinder block 1 off		29	Screw 4 off
3	"O" ring 2 off		30	Plain washer 4 off
4	Nut (12 point) 8 off		31	Cylinder head bolt 2 off
5	Tappet guide block 2 off		32	Cylinder head torque stay bolt 2 off
6	Tappet block screw 2 off		33	Oil feed bolt 2 off
7	Serrated washer 2 off		34	Copper washer 6 off
8	Tappet, racing type 4 off		35	Domed nut 2 off
9	Cylinder head c/w guides 1 off		36	Inspection cap 4 off
10	Valve guide 4 off		37	Fibre washer 4 off
11	Valve guide circlip 4 off		38	Rocker (right exhaust, left inlet) 2 off
12	Exhaust pipe adaptor 2 off		39	Rocker (left exhaust, right inlet) 2 off
13	Inlet valve 2 off		40	Rocker ball pin 4 off
14	Exhaust valve 2 off		41	Rocker adjusting pin 4 off
15	Bottom cup 4 off		42	Adjuster locknut 4 off
16	Inner valve spring 4 off		43	Thrust washer, 3/8 in. 4 off
17	Outer valve spring 4 off		44	Thrust washer, 7/16 in. 6 off
18	Top collar 4 off		45	Spring washer 2 off
19	Split cotter 8 off		46	Rocker spindle 2 off
20	Cylinder head bolt 4 off		47	Sealing rubber 2 off
21	Plain washer 8 off		48	Copper washer 2 off
22	Cylinder head gasket 1 off		49	Domed nut 2 off
23	Inlet rocker box 1 off		50	Pushrod 4 off
24	Exhaust rocker box 1 off		51	Cover tube 2 off
25	Joint washer 2 off		52	Rubber washer 4 off
26	Rocker box stud 4 off		53	Locking spring 4 off
27	Plain washer 4 off			

7 Dismantling the engine - removing the contact breaker and timing case

1 The contact breaker assembly is housed in the right hand side of the engine, under the chrome cover of the timing case. Remove two Phillips crosshead screws and detach the cover and gasket. Remove the two pillar bolts which hold the backplate to which points are attached, and let the backplate hang out of the way, suspended on the wires.

2 Now remove the advance and retard unit by withdrawing the centre bolt. A sharp tap with an alloy or bronze drift will usually release the taper and the unit will fall free. Alternatively, screw in a long 5/16 in UNF bolt and clench it with a Mole wrench, then tap the wrench with a hammer.

3 If a Triumph service tool is available - before engine number H57083, tool no D484, after engine number H57083, D782 - this should be used for extracting the contact breaker cam and advance/retard unit.

4 Pull the rubber boot off the oil pressure switch on the front right hand side of the crankcase (1968 models onward) and disconnect the spade connector.

5 Unscrew the eight crosshead screws of the timing cover with a good fitting screwdriver (an impact screwdriver is a worthwhile although expensive asset especially for crosshead screws). There may be a small amount of oil trapped in the case so take off the cover over a drip tray.

6 If the joint does not break easily, a few light taps with a soft headed hammer will help. Pull through the wires from the contact breaker points and remove the timing cover.

8 Dismantling the engine - removing the oil pump and timing pinions

1 The oil pump and timing pinions are now exposed. Remove the two domed nuts holding the oil pump and pull the pump off the studs, taking care not to lose the drive block. A thin paper gasket fits on the joint face and should be discarded.

2 To remove the camshaft pinions, unscrew the nuts holding them on the camshafts. Both have left hand threads. Note that it is not necessary to remove the camshaft nuts or cams to separate the crankcases and it is best to leave them in place. If the nuts have been removed, look down the end of the threads and note which of the three keyways line up with the key, then mark it. This is very important. The cam pinions are marked with dashes and/or dots for reassembly purposes.

3 Extractors are recommended for the removal of the timing pinions but they can be removed, with great care, by using two levers with hooked ends, placed opposite each other so that they will gradually ease off the pinion. Alternatively, Triumph tool No D2213/D2217 should be screwed onto the camwheel with the centre thread backed right off and when the body is on as far as the thread permits, the centre screw is used to extract the pinion. Repeat this procedure for the other pinion.

4 The crankshaft pinion nut has a right hand thread. If the engine persists in turning over, place a bar of a good fit through the small end eye of both connecting rods and let it rest on the crankcase top to lock the engine. Triumph tool No 61-6019 is advised to extract the pinion. Locate the three claws of the tool under the rear face of the pinion and tighten up the large outer collar. Use the middle thread as the extractor. If no special tool is available the lever method can again be used, although with extreme caution so as not to damage the gasket faces or the pinion itself.

5 The method of retaining the camshafts enables them to be removed and replaced without need to completely strip the engine. Refer to Section 27 of this Chapter for full details. This applies to engines from H 65573 onwards only.

6.4 Rocker boxes are retained by Phillips screws

6.6 Lift out pushrods after rocker box removal

6.7 Cylinder head is retained by four bolts

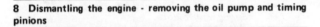

6.7a Pull off push rod tubes

6.8 A thin spanner is needed for the base nuts

6.9 Wedge cam followers with rubber to prevent displacement

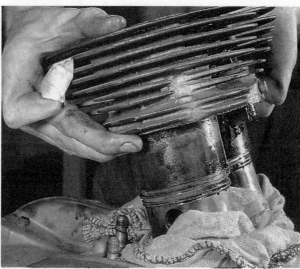

6.10 Pad crankcase with rag prior to pistons emerging from bores

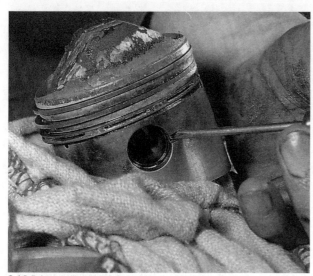

6.13 Prise out wire circlips and discard them

6.13a If gudgeon pin is tight, use alloy drift

7.1 Pillar bolts retain contact breaker base plate

7.2 Remove centre bolt to free advance unit

7.2a Use another bolt as means of extraction

7.6 Light tap should free timing cover

8.1 Oil pump drive block must not be mislaid

8.2 Both camwheel pinions have left-hand thread retaining nuts

8.2a Mark keyway in use before withdrawing pinion

8.4 Use of screwdrivers to extract crankshaft pinion

8.4a Retaining nut has a right-hand thread

9 Dismantling the engine - removing the primary drive

1 Place a drain tray under the primary chaincase and remove the drain plug located at the bottom rear end. Unscrew the chain tensioner completely and remove it.

2 Unfasten the crosshead screws around the primary chaincase. Withdraw the cover and paper gasket and remove the chain tensioner assembly.

3 Unfasten the three nuts holding the stator and rest the stator out of the way.

4 Bend back the tab washer holding the centre nut on the rotor. Using a bar through the connecting rods as before to lock the engine, and remove the rotor retaining nut. Pull off the rotor and place it away from any metal as it is a very strong magnet.

5 As the primary chain is endless, both the engine sprocket and the clutch must be removed as one. Slacken and remove the three or four (as applicable) screws in the end of the clutch using a penknife placed behind each screw head to depress the spring whilst the screw is slackened and removed. The screws have location nibs at their rear to prevent them from slackening when the engine is run.

6 Withdraw the cups and springs. The individual clutch plates, both plain and bonded, can now be hooked out using two short pieces of wire bent at one end.

7 When all the clutch plates are withdrawn, access is available to the inner drum and shock absorber unit. With the engine still locked in position, and a locking bar to scotch the clutch, unscrew the centre retaining nut and remove it together with the

cup washer. (It should be noted that machines before H.49833 have a tab washer which must be bent back first.)

8 The complete clutch assembly can now be pulled off the centre by inserting Triumph service tool D662/3 and screwing it home fully so that the full depth of thread engages. When the centre bolt is tightened the clutch will be drawn off the mainshaft taper. For machines with engine numbers before H.49833 use tool D/A50/1.

9 If a service tool is not available, the clutch pressure plate can be used to equally good effect. Refit it without the clutch plates or the cups and springs using the adjuster nuts and washers placed beneath them, as shown in the accompanying photograph. It will be necessary to file a pip in each washer to accommodate the pip on the underside of each adjuster screw. Unscrew the adjuster back nut in the centre of the clutch pressure plate and screw the adjuster inwards. It should then pull the clutch assembly off the splines in the clutch centre by the centre screw exerting pressure on the end of the mainshaft.

10 The above technique should not be used if the clutch is an extremely tight fit on the spline. In this case the Triumph service tool must be used.

11 Before the primary drive can be released it is necessary to withdraw the engine sprocket. Triumph service tool 61-6-46 is recommended. The extractor bolts screw into the sprocket and when the centre nut is tightened it will draw off the sprocket. If the tool is not available there is room for a two or three legged sprocket puller as an alternative means of extraction. Note the disposition of the rotor, engine sprocket and any spacer(s) fitted as an aid to correct reassembly. It is important that these components are reassembled in the correct order.

12 If the Triumph service tool is not used there will be a deluge of rollers falling from the back of the clutch as soon as it is removed. If any are lost there is a total of 20 rollers.

13 If the ends of the sprocket puller legs are too thick to go behind the clutch hub that remains, end float can be easily obtained by removing the outer gearbox cover and releasing the nut on the opposite end of the mainshaft.

14 Unfasten the clip on the underside of the crankcase which holds the alternator wires and the tube nut inside the chaincase through which they protrude, and pull through the wires so that the alternator stator can be lifted away.

10 Dismantling the engine - separating the crankcases

1 The crankcases can now be separated by removing the various nuts and bolts which hold them together. There are two screws inside the crankcase mouth which it is recommended should be taken out first, and a nut behind the outer gearbox cover. The outer gearbox cover must be removed to gain access to it. See Chapter 2.

2 Pull the crankcases apart to release the crankshaft assembly,

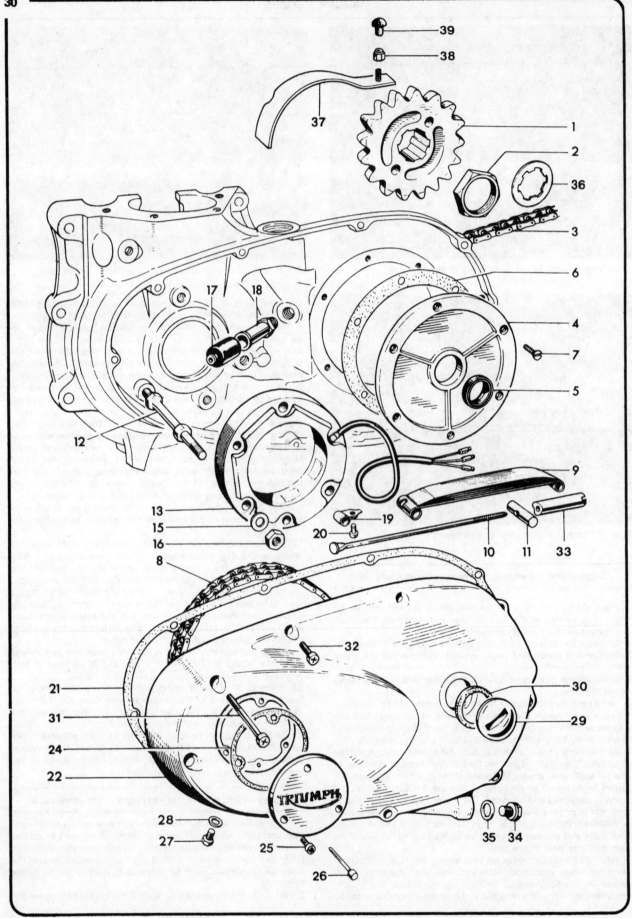

Fig. 1.2. Chaincase and chains

1	Gearbox sprocket (17 teeth) 1 off
1	Gearbox sprocket (18 teeth) available
1	Gearbox sprocket (19 teeth) available
1	Gearbox sprocket (20 teeth) available
2	Gearbox sprocket nut 1 off
3	Rear chain, 5/8 in. x 3/8 in. x 102L 1 off
3	Rear chain, 5/8 in. x 3/8 in. x 103L 1 off
4	Cover plate 1 off
5	Oil seal 1 off
6	Joint washer 1 off
7	Countersunk screw 6 off
8	Primary chain, 3/8 in. Duplex 78L 1 off
9	Chain tensioner blade 1 off
10	Tie rod 1 off
11	Trunnion 1 off
12	Stud 3 off
13	Stator, RM21 (47205) 1 off
14	Serrated washer 1 off
15	Plain washer 3 off
16	Nut
17	Rubber grommet 2 off
18	Sleeve nut 1 off
19	Clip 1 off
20	Setscrew 1 off
21	Joint washer 1 off
22	Chaincase 1 off
23	Rotor cover 1 off
24	Joint washer 1 off
25	Screw 3 off
26	Ignition pointer 1 off
27	Level plug 1 off
28	Copper washer 2 off
29	Filler plug 1 off
30	Fibre washer 1 off
31	Screw 3 off
32	Screw 7 off
33	Adjuster sleeve nut 1 off
34	Drain plug 1 off
35	Fibre washer 1 off
36	Tab washer 1 off
37	Crankcase protector 1 off
38	Self-locking nut 1 off
39	Grommet 1 off

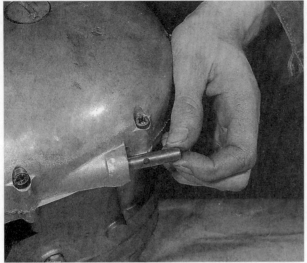

9.1 Unscrew chain tensioner completely

9.3 Stator assembly will pull off studs

9.4 Rotor nut is very tight (R/H thread)

9.6 Lift out clutch plates

9.7 Metal strip used to scotch inner drum

9.9 Clutch pressure plate used to extract clutch centre

taking care whilst doing so that the timed breather disc and spring on the end of the inlet camshaft does not get lost. The left hand crankcase has a circular plate which covers the final drive gearbox sprocket, but there is no need to remove this unless it is leaking or damaged. To remove it, in order to replace the oil seal or gasket, undo the six countersunk screws and push the plate out from the rear. Renew the gasket or oil seal, clean off and refit.

Prior to engine No H65573

3   The left hand crankcase contains a ball journal bearing which must be inspected for wear. By heating up the crankcase, the bearing housing will expand and a sharp tap will drive the bearing inwards to permit inspection. The oil seal behind the bearing should be tapped outward and a new replacement fitted.

4   The right hand crankcase contains a large bearing bush which is removed by undoing the locking plate on the inside of the timing case. Heat the crankcase and tap the bush out by using a nylon drift or similar soft instrument.

After engine No H65573

5   The left hand crankcase has a roller bearing, the centre of which will remain on the crankshaft. This roller cage and the outer track retained in the crankcase should not be disturbed unless the bearing has to be renewed. To remove the outer race, heat the crankcase around the bearing housing. When hot, put a wet rag rolled into a ball in the centre of the race and tap the crankcase downward on a flat piece of soft wood to shock the race out of position. Be very careful not to damage the crankcase sealing face or to hit it too hard. Triumph service tool 61-6060 is available to simplify this task. Heat the crankcase, insert and expand the service tool, then tap it out.

6   The right hand crankcase has a ball journal bearing. To remove this bearing, heat the crankcase and tap the bearing inward. This will be necessary only if a replacement is needed.

11 Examination and renovation - general

1   With the engine completely stripped down, examination for wear can now take place. First clean all the parts thoroughly. It is always a good idea to have sheets of paper laid out on which the parts can dry without fear of them getting dirty again.

2   Examine all the castings for cracks or any other signs of damage, ie damage to any jointing surface or bearing surface, and where alloy castings butt against each other.

3   Should any studs or threads need attention, now is the time. Care is necessary when removing or replacing them. Castings are sometimes weak at these points. Beware of shearing them off or overtightening which will cause more problems.

4   Where internal threads are stripped or badly worn, it is preferable to use a thread insert rather than tap oversize which weakens castings and sometimes does more harm than good. Most dealers can provide a thread reclaiming service by the use of a Helicoil insert, which is better than the original in aluminium castings.

12 Main bearings and oil seals - examination and renovation

1   When the main bearings have been removed, wash them in a petrol/paraffin mix and inspect for wear or pitted tracks. If there is any play in a ball race or if it runs roughly, then renewal is necessary. This also applies to roller bearings. Should the bearings be loose on their shafts or in their housings the use of Loctite Bearing Fix is recommended.

2   Oil seals must always be replaced when the engine is being rebuilt, even if no visible sign of wear is apparent.

13 Crankshaft assembly - examination and renovation

1   Clean the crankshaft completely. If the connecting rods are

9.11 Use of a two-leg puller to withdraw engine sprocket

9.12 Clutch bearing rollers are uncaged

9.14 Clip under engine for stator wire

10.1 Screws within crankcase mouth must be removed first

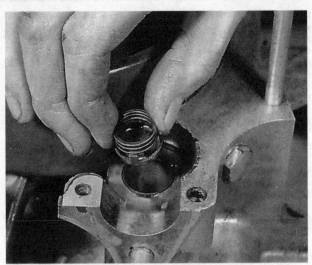

10.2 Don't lose breather disc and spring

10.2a Circular cover provides access to final drive sprocket

10.4 Bearing bush is retained by locking plate

not marked, mark them to distinguish the side, front, back, etc so that they are not interchanged. Keep the nuts and bolts together.

2   Inspect the bearing surfaces for wear, which usually takes the form of scoring or scuffing. It is advisable when you have got this far to renew the big end shells - which are reasonable cheap, provided the bearing surfaces are good.

3   Specialist attention is required if the crankshaft bearing surfaces are scored or damaged in any way. Under no circumstances must emery tape or any abrasive be used to erase even one small mark. If the crankshaft is worn or scored get it reground or purchase a service exchange crankshaft. Remember, this is the heart of the engine and scrimping now will have catastrophic results later.

4   Journals or bearing surfaces can be reground as can the timing side bushes, where fitted. When a crank has been ground the term undersize will be used, either -010 in or -020 in, because grinding reduces the diameter of the journal. Bearings will then be supplied to suit.

5   The size of undersize of bearings will be marked on the back of each shell. Under no circumstances should the white metal face be scraped or damaged. Even a touch with the finger nail can destroy the finish, so great care must be taken whilst handling and fitting them.

6   It is strongly recommended that the connecting rod bolts should be renewed, as they are highly stressed. If one should happen to break whilst the engine is running, it will cause a total write-off. Fit new bolts and tighten to the correct torque setting or extension when fitting.

7   It is not usual for the crankshaft to be stripped, but if in doubt about the oil supply to the bearings, the centrifugal sludge trap which is incorporated should be cleaned out. Remove the large headed screwed plug on the right hand side of the crankshaft and the bolt adjacent to the journal, and with the aid of a piece of wire, pull out the tube. Clean the tube thoroughly and wash it out, then replace it so that it locates with the bolt before screwing in the plug.

8   If the crankshaft has been reground it is advisable to remove and clean the sludge trap to clean out any grinding dust which may have found its way in.

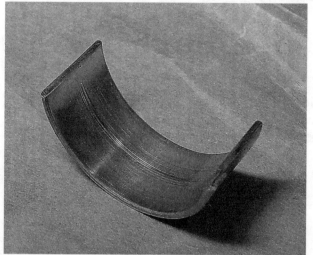

13.2 Scored bearing shells must be renewed

13.3 Inspect crankshaft bearing surfaces at same time

14 Camshaft and timing pinion bushes - examination and renovation

1   Unless the machine has covered an extremely high mileage, no attention should be necessary.
2   The camshaft bushes (there are only two; the timing side of each camshaft runs directly in the crankcase casting) are found in the left hand crankcase half. The rear one is pinned and also houses the timed breather on some models. To remove the bushes, heat the crankcase and use a lipped punch behind to extract. Refit the new bushes whilst the crankcase is still hot, making sure that any oil holes line up with those in the crankcase.
3   If the exhaust camshaft has a rev counter drive attached to the end, remove it first.
4   The bushes are made from sintered bronze and may require the light use of a reamer to make them fit perfectly. Measure the camshaft bearing surface and ream until a good running fit is achieved.
5   The idler pinion bush, if worn, can be tapped out and a new one tapped in with a soft material to avoid damaging it in any way. Use an adjustable reamer if necessary and ream to a good running fit as before.

15 Camshaft, cam followers and timing pinions - examination and renovation

1   The camshafts can be inspected best under a strong light. If signs of scuffing and dipping are evident, look at the cam followers, which, when worn, dip and scuff badly in the maximum load area, ie the centre of radius. If they are badly worn they should be renewed. If they are only lightly scuffed, they can be cleaned up with a fine oil stone.
2   Check each timing pinion for worn or broken teeth. If they are badly worn then renewal is recommended. Do not be misled by noisy timing gears which can be attributed to backlash.

16 Cylinder barrel - examination and renovation

1   The depth of the lip at the top of each cylinder bore denotes how much the cylinder barrel has worn. Scores or grooves vertically up the bore also denote wear. If heavy scoring has taken place then a rebore is necessary. If the bores look good it is probable that only new piston rings will be necessary.
2   To check the piston rings ease them off the piston and press one of them down into the bore with the piston (be careful as they are very brittle and easy to break). Check the end gap

with a feeler gauge; if more than 0.012 inch replace all the rings.
3   The barrel can be rebored to + .040 inch. If your machine has had a rebore it will be marked on top of the piston - either +0.010 inch or +0.020 inch or +0.040 inch. If none of these is stamped then it can be assumed that the bore is standard.
4   Check that the base of the cylinder barrel is not cracked or broken. If it is, a new barrel must be fitted. This also applies if the maximum size is already +0.040 inch or the bores are badly worn or scored.
5   Cleaning off all road dirt and paint the fins either matt or 'pot' black to help the heat distribution. It also adds to the well finished look when the engine is reassembled.

17 Piston and piston rings - examination and renovation

1   The pistons and rings can be overlooked if the cylinder barrel is having a rebore because new pistons and rings will be fitted in the appropriate oversize.
2   If a rebore is not required it does not necessarily follow that the pistons are in good shape. Check for scores, cracks and damage to circlip grooves and renew if necessary. If the pistons are usable clean off each crown and polish them. Carbon does not adhere so readily to a polished surface.
3   Use an old piston ring to scrape carefully around the ring grooves and release any build up of carbon.

18 Small end bushes - examination and renovation

1   The play in a small end bush can be ascertained by passing the gudgeon pin through the bush - the pin should be a nice sliding fit.
2   If renewal is necessary use the simple draw bolt method as illustrated. Take great care to line up the oil hole in the bush with the hole in the top of the connecting rod.
3   When fitted, it is usually necessary to ream the bush. If the engine is still in the frame put rag around the mouth of the crankcase and ream until the gudgeon pin is a good sliding fit.

19 Cylinder head and valves - dismantling, examination and renovation

1   It is best to remove carbon deposits from the cylinder head before taking out the valves using a piece of soft material to avoid scratching the combustion chambers. Finish with metal polish. It is also advisable to screw in a set of old spark plugs to keep the threads clear.
2   The valves must be removed with a valve spring compressor.

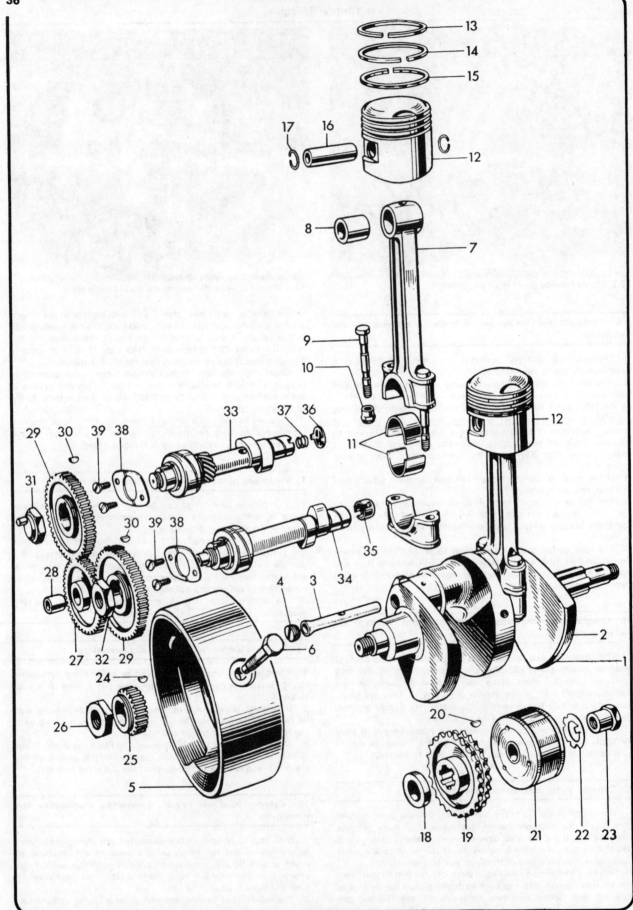

Fig. 1.3. Crankshaft and connecting rods (From engine No. H 65573)

1	Crankshaft and flywheel assembly 1 off	17	Circlip 4 off
2	Crankshaft 1 off	18	Distance collar 1 off
3	Oil tube 1 off	19	Engine sprocket, 26T Duplex 1 off
4	Screwed plug 1 off	20	Rotor key 1 off
5	Flywheel 1 off	21	Rotor, type RM21 (54213901)1 off
6	Flywheel bolt 3 off	22	Tab washer 1 off
7	Connecting rod 2 off	23	Nut 1 off
7	Connecting rod 2 off	24	Key 1 off
7	Connecting rod 2 off	25	Timing pinion 1 off
8	Small end bush 2 off	26	Nut 1 off
8	Small end bush 2 off	27	Intermediate wheel 1 off
9	Connecting rod bolt 2 off	28	Bush 1 off
10	Self locking nut 2 off	29	Camshaft wheel 2 off
11	Big end bearing (available in −.010 & −.020 undersizes) 4 off	30	Camshaft key 2 off
		31	Inlet camshaft nut 1 off
12	Piston complete CR 9.5 : 1 2 off	32	Exhaust camshaft nut 1 off
12	Piston complete CR 9.1 2 off	33	Inlet camshaft 1 off
13	Plain compression ring 2 off	34	Exhaust camshaft 1 off
13	Taper compression ring 4 off	35	Tachometer drive thimble 1 off
14	Taper compression ring 2 off	36	Rotary breather valve 1 off
15	Oil control ring 2 off	37	Breather valve spring 1 off
15	Oil control ring 2 off	38	Retaining plate 2 off
16	Gudgeon pin 2 off	39	Screw 4 off

The compressor compresses the spring and releases the collets, which are held to the top of the valve stem by a taper in the spring cap.

3   Take out one valve at a time, and before grinding in, check the fit between the valve and the valve guide, which is in the head. Always replace the valve in their correct positions.

4   If the valve guide play is not excessive then regrind the valve. First clean the back of the valve, taking care not to damage the seat. Check each valve for damage to its seat and the seat itself for pocketing which means that the valve is sinking back into the head. If this has occurred get it re-cut at your dealers.

5   Using a small rubber suction pad on a stick, and some carborundum paste, apply the sucker to the valve head and a small amount of paste to the valve seat. Spin the grinding tool between the hands and every three or four spins, lift away from the seat; if a little paraffin is put on the valve stem and on the paste it will cut in more easily. The object of this is to re-align the valve and seat and to eradicate any imperfections in it. A grey, dull colour shows when the seating is restored (on both valve and seat).

6   If there is any doubt about the valves, renew them - the large diameter valves are the inlet, the small ones are the exhaust.

7   If the valve guides are badly worn replacements are recommended. Remember that a quite large tolerance can be expected because of the extreme heat the cylinder head experiences.

8   To replace the valve guides, heat the cylinder head until it is very hot and tap out the guides from the inside. Some guides are bronze and others cast iron. A valve guide extractor can be quite easily fabricated from a 5 inch length of mild steel bar turned to 5/16 inch diameter for a length of 1 inch.

9   The circlips should be removed with the valve guides and refitted when tapping in a new guide. When the circlip is right against the head and in its correct position, a different note in the tapping will occur.

10 If the machine is used for high speeds it is recommended that new valves be fitted as a matter of course. Valves suffer from fatigue more than most parts of the engine.

11 The valve springs should be renewed as a matter of course. They are reasonably cheap and the advantages overcome the cost. Good valve springs are a must for high performance, and make the whole engine run more efficiently.

12 Reassemble the valves in the reverse order of stripping, making sure to oil liberally all moving parts with clean, new oil.

13 Finally make sure that the gasket face of the head is flat, by lapping on an old sheet of glass. (An old car window is perfect, providing it is flat.) Put fine wet and dry rubbing down paper on it and tape down around the edge, then with a rotary motion, rub the head without pressing hard, until a grey matt finish on the jointing surface is achieved.

14 Wash off the surface and dry. Remove the old spark plugs, making sure that no carbon is left in the threads. Wrap the cylinder head in rag and put it aside. The cylinder head is now ready for reassembly.

20 Cam followers and cam follower guide blocks - examination and renovation

1   The cam followers and guide blocks usually need no attention but a check for wear is necessary in case the machine has been run low in oil, or some similar catastrophy has occurred. Rock the follower in the guide block. It should be a good sliding fit, with very little sideways movement.

2   If excessive wear is found remove the cam follower block and the small location screws on the front and back of the barrel adjacent to the blocks. Use Triumph service tool 61-6008 to drift them out, upwards, taking care not to hit the drift too hard. Cylinder barrels have been known to break by over-zealous hammering during this operation.

3   To replace them use the same drift and after renewing the O ring, grease the block, lightly line up the location hole and drift it in until right home. Repeat this for the other block; refit the locating screws and tighten them.

19.2 Use a valve compressor to free valve collets

19.3 Check valve guides for wear

19.11 Renew valve springs every top overhaul

4 If the cam followers need replacing, they must be replaced in their correct positions, otherwise poor lubrication and damage is likely to occur.

5 Remember that as soon as the engine is started on reassembly all cam and valve gear is running under extreme pressure, so oil all components liberally with clean engine oil.

21 Removing and replacing the tachometer drive box

1 Where a tachometer is fitted the drive is taken from the exhaust camshaft left-hand end. On 1964 and 1965 models the cable is connected directly to a crankcase-mounted adaptor, the drive being transmitted by a spade from the camshaft. On all later models the cable is connected to a separate gearbox which turns the drive through 90° and gives a 2:1 reduction ratio; this complete assembly can be fitted to earlier models.

2 To remove the gearbox, disconnect the cable and unscrew the large slotted end cap. Extract the drive pinion, using either long-nosed pliers or by turning the engine over quickly. This will reveal the central retaining sleeve bolt, which must be unscrewed using a slim box spanner or deep socket. Note that on 1966 to 1968 models a 3/8 in Whitworth spanner will be required and the bolt uses a normal right-hand thread. On all later models (engine no. H65573 on) a 7/1 6 in AF (approx 3/16 in Whitworth) spanner will be required and the bolt uses a left-hand thread, ie it is unscrewed clockwise. Withdraw the gearbox, noting the special sealing washer (stat-o-seal) fitted as standard to 1973 on models; this is fitted to prevent slackening as well as preventing oil leakage, and must be renewed whenever the gearbox is disturbed.

3 To remove the driven gear, first unscrew the locating set screw (early models) or drive out the locking pin (later models) and withdraw the driven gear housing, noting the sealing O-ring. Note that the housing will be a tight fit. The driven gear can then be withdrawn. On reassembly renew all O-rings and seals.

4 On early models a slotted thimble is pressed into the exhaust camshaft end to transmit drive to the spade; if damaged the thimble can be renewed from outside the crankcases, providing care is taken to extract as much of the old part as is possible, before carefully locating the new part so that the drive spade can engage it fully. Note that a modified drive thimble was introduced in late 1966, which was of strengthened design and smaller in diameter to ease fitting; ensure that this type is used as a replacement. On later models (1970 on) a slotted plug is threaded into the camshaft end. Note that this is not interchangeable with the earlier type.

5 Reassembly is a direct reversal of the removal procedure, noting that all O-rings and oil seals should be removed as a matter of course.

22 Pushrods, rocker spindles and rocker arms - examination and renovation

1 Check the pushrods for straightness. If bent, they cannot be renovated and therefore must be replaced.

2 Check the cups on the end of the pushrods for tightness. If loose. Bearing Fit Loctite will take up a large amount of slack, but if the wear is too great, the pushrod should be renewed.

3 Examine the rocker arms and renew the ball end if it is badly scored, also the adjuster if it is worn. Before engine No H.65573 locate the flat towards the spindle when replacing the ball ends.

4 Wear in the rocker arm spindles is very rare, unless they have been starved of oil or an extremely high mileage has been covered. To replace, use a soft drift to drive out the spindle and remove the rocker arms and washers, taking note of the location of the various washers.

5 Clean off all the components thoroughly and blow through all the oilways. Remove the oil seals from the spindles and renew them all.

21.2 Remove large slotted end cap to give access to tachometer gearbox retaining bolt

21.5 Press gear into position to relocate

6 It is a good idea, when at this stage, to lap in the cylinder head barrel joint, after just removing the two remaining studs by locking together two nuts to extract them. Lap with a sheet of plate glass and a sheet of very fine wet and dry rubbing down paper, using a rotary motion until a dull, grey consistent finish is achieved over all. Clean thoroughly, then replace the studs.

7 Before commencing assembly of each rocker box, note that there is one plain washer in each box which has a smaller diameter hole. This is the thrust washer through which the smaller diameter of the spindle enters. This is assembled last, against the left hand inner face of each rocker box.

8 With the rocker box resting base uppermost and using grease to hold the components in place, start from the right and with the spindle just in the box, slide on the first large-holed flat washer, then, moving the spindle in a little at a time, add the other components as per Fig. 1.7. Repeat this for the other rocker box. Oil liberally.

23 Engine reassembly - general

1 Make sure that every component is clean and that all traces of old gasket of old gasket have been removed.

2 Make sure that the work area is clean. It is also a good idea to have either brown paper or newspaper to cover the bench.

3 Your tool kit should be clean and the right size for the job.

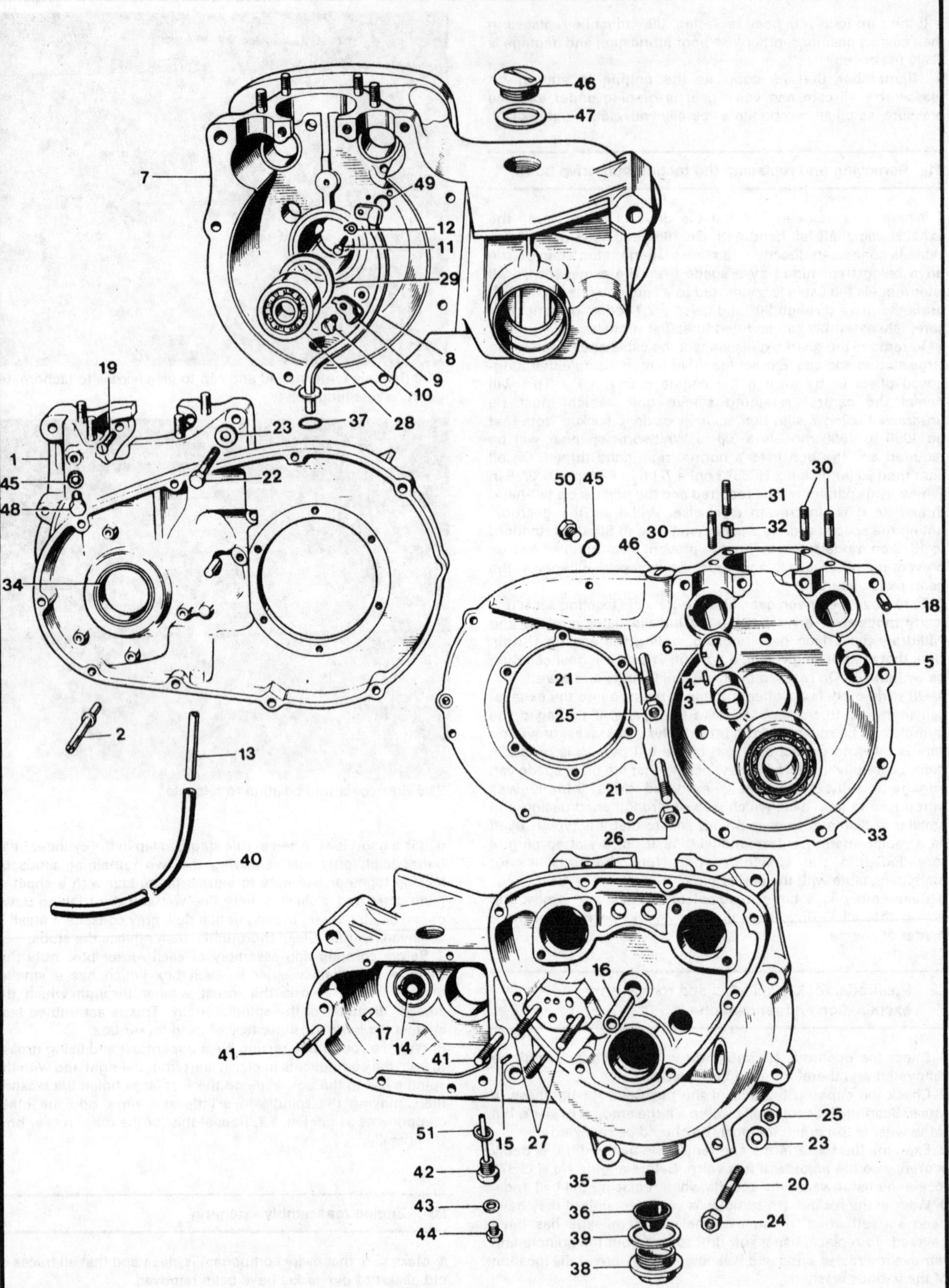
46
47
7
49
12
11
29
8
9
10
28
37
19
23
1
22
45
48
34
2
13
50
45
30
31
32
46
18
6
5
21
4
3
25
21
33
26
40
16
17
14
41
41
51
15
27
25
23
42
35
20
43
36
24
39
44
38

Fig. 1.4. Crankcases

1	Crankcase drive side 1 off		27	Oil pump stud 2 off
2	Pivot pin 1 off		28	Right main bearing (ball journal) 1 off
3	Inlet camshaft bush 1 off		29	Abutment ring 1 off
4	Locating pin 1 off		30	Cylinder base stud 6 off
5	Bush exhaust camshaft 2 off		31	Cylinder base stud 2 off
6	Rotary valve 1 off		32	Dowel 2 off
7	Crankcase timing side 1 off		33	Left main bearing (roller journal) 1 off
8	Oil scavenge pipe 1 off		34	Oil seal 1 off
9	Clip 1 off		35	Filter spring 1 off
10	Screw 1 off		36	Filter 1 off
11	Bolt 1 off		37	Sealing washer 1 off
12	Tab washer 1 off		38	Filter cap 1 off
13	Breather pipe 1 off		39	Joint washer 1 off
14	Needle roller bearing 1 off		40	Breather extension pipe 1 off
15	Dowel at junction block		41	Stud 2 off
16	Idler gear spindle 1 off		42	Drain plug (with level tube) 1 off
17	Peg (at thrust washer) 1 off		43	Fibre washer 1 off
18	Hollow dowel 2 off		44	Level plug 1 off
19	Screw 2 off		45	Fibre washer 1 off
20	Stud 1 off		46	Filler plug 2 off
21	Stud 2 off		47	Fibre washer 1 off
22	Bolt 1 off		48	Blanking plug (LH thread) 1 off
23	Plain washer 3 off		49	"O" ring 1 off
24	Nut 3 off		50	Plug 1 off
25	Nut 1 off		51	Washer 1 off
26	Nut 1 off			

Screwdrivers should have keen, sharp edges. You will also need a hide mallet, a few alloy drifts and a torque wrench. There is nothing worse than having to stop half way through owing to lack of tools.

4   Finally, you will require new oil seals and gaskets and clean engine oil for the finished engine, also some clean engine oil in an oil can. A well-lit shed, garage, or even kitchen is the other requirement, then you are ready for your engine rebuild.

24   Engine reassembly - reassembling the crankshaft

1   Ensure that the connecting rod caps and connecting rods are scrupulously clean. Wash the preservant oil from the big end shells with clean petrol.

2   Place the big end shells in the caps and rods, locating the tabs on the shells. **Do not scratch the surface of new shells.** Smear with clean oil and assemble, making sure that the caps are mated with the marks made previously on the rods.

3   With the new nuts and bolts in place, tighten to 18 lb ft (2.489 kg m) with a torque wrench, or if no torque wrench is available, then measure with a micrometer to a bolt stretch of 0.005 inch.

4   Pump oil from oil can through the crankshaft and turn the connecting rods at the same time until the oil appears around the journals.

24.2a All parts should be replaced in their original order

25   Reassembling the crankcase

1   If the main bearings have been removed they should now be replaced. Before engine No H.65573 the timing side bush should be tapped in after the crankcase has been heated, and then the screw and locking plate put back with Loctite.

2   The drive side crankcase has a ball journal bearing and this should be tapped in after heating the crankcase. The oil seal can be put in from the outer side, when the case cools down.

3   After engine No H.65573 a roller bearing is fitted in the left hand drive crankcase. The crankcase should be heated and the outer cage dropped in with the lipped side of the cage toward the oil seal housing. Allow to cool down and fit the oil seal. The centre of the roller cage must now be heated in very hot oil and tapped on the drive side of the crankshaft.

4   The right hand crankcase has a ball race. This is also fitted whilst the crankcase is warm.

5   Make sure that the crankcase joining surfaces are clean. Apply a coat of jointing compound. With the left hand drive crankcase held firmly, place breather disc and spring in the inlet cam bearing the right way up so that the tabs locate with the end of the camshaft.

6   On machines before engine No H.65573 put the crankshaft into the drive side first, and then lower on the other half and tap together, being very careful not to trap the connecting rods. If trouble occurs with the breather disc, turn the inlet camshaft to locate it in the slots.

7   Machines with engine numbers after H.65573 are assembled differently. The timing side is tapped onto the crankshaft first, followed by lowering the assembly into the other half. If the breahter disc will not locate, turn the inlet camshaft until you feel it click in. Tap the cases together and fit the screws and nuts.

8   Torque load crankcase junction bolts to 15 lb ft (2.07 kg m) and junction studs to 20 lb ft (2.765 kg m). Do not force the crankcases together. If they do not go together when tightened there must be a reason - find and rectify it. **Do not forget the two screws inside the top of the crankcase and make sure that they are tightened fully. Use Loctite.**

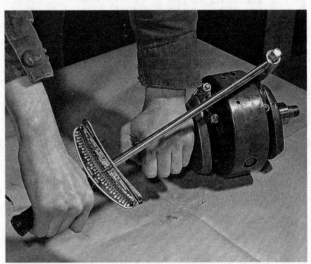
24.3 Use torque wrench for correct loading of nuts

26   Engine reassembly - replacing the timing pinions, oil pump and timing cover

1   With the camshafts in position and the crankcases together,

25.1 Screw in locating plate must be tight

replace the washer on the end of the crankshaft with the chamfer outward.

2   Replace the Woodruff key in the crankshaft (if not a good fit, renew it) and with the timing marks to the outside, align the crankshaft pinion and drift on.

3   Using the keyways marked when stripping the engine, line up each camwheel and drift into position on the respective camshaft. On single carburettor models, align the dots on the camshaft pinion so that they coincide exactly with the dots on the idler pinion as it is pushed into position. Retain it in place with a rubber band. (Do not forget to remove the band before fitting the outer case.)

5   On twin carburettor machines use the dash marks of the camshaft pinions for timing, as shown in the accompanying illustration.

6   Lock the engine with a bar through the small end eyes of the connecting rods or if the engine is still in the frame, engage gear and put on the rear brake.

7   Spin on and tighten the cam nuts, not forgetting the one with the oil pump drive pin goes on the inlet camshaft. **Both nuts have left hand threads.**

8   Replace and tighten the engine pinion nut.

9   Flood the oil pump with oil and put the drive block in the top of the pump, line up and push into place. Tighten the nuts and use Loctite. If possible torque to 6 lb ft (0.83 kg m). **Do not use jointing compound on the oil pump joint gasket, which must be renewed.**

10 The timing cover can now be fitted; if new oil seals are necessary fit them now. The crankshaft engine seal is held in with a circlip - to renew it, remove the circlip, hook out the old seal, fit the replacement and replace the circlip. The camshaft oil seal will press out of position.

27   Removing and replacing the camshafts only, without separating the crankcases

From engine No H65573 only

1   It is not necessary to separate the crankcases in order to replace the camshafts.

2   Remove the rocker boxes and the timing cover. Take out the oil pump by taking off the two conical nuts and locking washers. Block the relevant holes in the crankcase face to avoid any loss of engine oil, but remember to unblock them when the oil pump is refitted. Take off the inlet and exhaust camwheels, noting which keyways are in use. The camshaft retaining plates and screws will then be visible.

3   The retaining plate screws are centre punched to lock them. Use a small drill to remove the centre punch marks, then extract the screws followed by the retaining plates. Tilt the machine over onto the left hand side, to stop the cam followers dropping into the crankcase. The camshafts can now be drawn out from the right hand side of the crankcase.

4   Oil the bearing surfaces and refit the camshafts and retaining plates, using new screws which should be centre punched. Reassemble in reverse order of removal.

5   Before engine number H.65573 machines were fitted with a breather disc and spring which was incorporated behind the inlet camshaft.

6   Care must be taken when removing the camshaft not to drop the disc and spring into the engine. When a new camshaft is fitted, place the rotary breather valve and spring into the camshaft bush. Assemble the camshafts making sure that the slot in the end of the inlet camshaft engages with the projection on the breather disc valve.

7   The timing case should now be replaced. Ensure that the jointing faces are clean and apply a fresh coat of jointing compound. Screw in the tapered oil seal pilot which is a necessity to guide the oil seal over the camshaft step. Use Triumph service tool D1810, if available, and grease lightly to assist assembly. Slide the cover into position and replace the eight crosshead screws. Tighten them fully.

25.5 Insert breather disc and spring first

25.6 Lower right-hand crankcase into position

25.8 Tighten and Loctite screws in crankcase mouth

28 Replacing the oil feed pipes, the gearbox end cover and the advance/retard unit

1 The oil feed pipes are now very easy to put on. Fit a new gasket and push them into place. Tighten the nut, not forgetting the washer.

2 Replace the gearbox cover as directed in the next Chapter, not forgetting the timing necessary.

3 Release the oil seal guide on the exhaust cam and locate the advance and retard unit on the peg. Insert the centre bolt and tighten.

4 Replace the contact breaker plate. Feed the wires through first; leave the pillar bolts loose as the machine must be timed later.

5 Screw on the chrome cover to protect the points.

29 Replacing the pistons and cylinder barrels

1 Oil the gudgeon pin holes and place a piston over the connecting rod on the side from which it came. Position it the right way round, with one new circlip in place. Tap in the gudgeon pin and insert the other circlip. **Double check that the circlips are in correctly.**

2 Repeat for the other piston. Put a layer of jointing compound and a gasket on the cylinder barrel/crankcase joint. Insert the cam followers in the guide blocks and hold them in place with rubber bands.

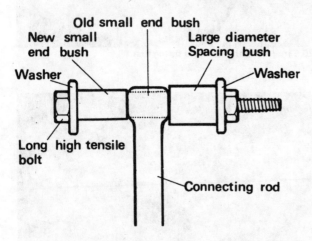

Fig. 1.5. Drawbolt method of replacing the gudgeon pin bush

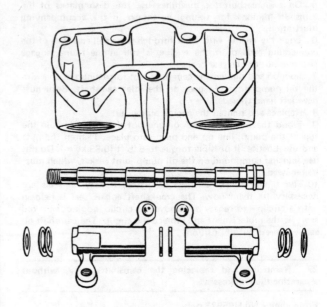

Fig. 1.7. Assembly of rocker arms

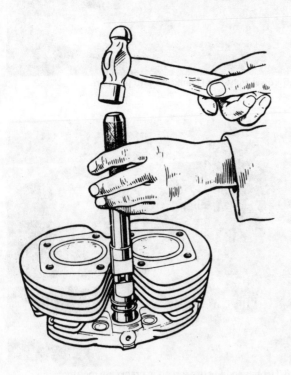

Fig. 1.6. Refitting a tappet guide block

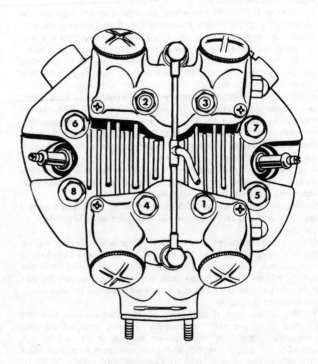

Fig. 1.8. Cylinder head bolt tightening sequence

26.2 Replace pinion with timing marks outwards

26.5 Timing pinions aligned correctly

26.7 Oil pump drive is on inlet camshaft pinion nut

26.9 Use new gasket when refitting oil pump

27.2 Keyway in use must be marked to aid reassembly

27.3 Retaining plate is held by two countersunk screws

27.3a Lift out only when engine is tilted to left

27.4 Centre punch retaining screws when tight

28.3 Advance unit will locate in only one position

28.4 Pillar bolts must have seating washers

29.1 Check circlip location very carefully

29.6 Piston ring clamps aid refitting of pistons

3 New piston rings should be already fitted. Oil the pistons and bores liberally.

4 Put a piston ring clamp on each piston and rest both pistons on bars across the mouth of the crankcase to steady them.

5 If piston ring clamps are not available, extreme care must be taken - with two people involved. One should lower the cylinder barrel whilst the other eases in one ring at a time on each piston.

6 Ease the cylinder barrel down all the way, remove the piston ring clamps (if used) and pull out the support bars (if used). Push the cylinder barrel into place and do up all the holding-down nuts, making sure that the barrel locates on the two dowels of the crankcase.

7 Check the engine rotates freely before proceeding further.

30 Replacing the pushrod tubes and cylinder head

1 Ensure that the cylinder head jointing surfaces are clean and grease lightly, then place on the new gasket.

2 Position the pushrod tubes and fit new seals (and cup on earlier models).

3 Lower the cylinder head after placing the top seals on the pushrod tubes, then locate the bolts finger tight.

4 Line up the pushrod tubes.

31 Replacing the pushrods and rocker boxes

1 Place the pushrods in their original positions, making sure that they are located correctly in the cups.

2 Carefully rotate the engine until the two inlet pushrods are down together.

3 Position the new rocker box gaskets and lower the rocker box, ensuring that the pushrods engage correctly with the rocker arms.

4 Insert the two bolts with their washers until they are finger tight.

5 Screw in the two outer Phillips screws (some late machines are fitted with bolts at this point). Replace the nuts and washers on the underside of the rocker boxes. Tighten them until they are just tight.

6 Tighten the cylinder head bolts in the correct sequence, as shown in the accompanying diagram. Tighten to a torque wrench setting of 18 lb ft (2.489 kg m) for the 3/8 inch bolts and to 5 lb ft (0.69 kg m) for the small nuts and bolts of the rocker box.

32 Engine reassembly - replacing the clutch, engine sprocket and primary chain

1 The primary chain fitted to the unit construction twins is of the duplex type. No spring link is fitted; the chain must be fitted together with the engine sprocket and clutch, as one and the same time. If the chain tensioner has been removed, it should be refitted now, slackened off.

2 Replace the clutch centre complete with roller bearings and clutch chainwheel after the Woodruff key has been inserted in the gearbox mainshaft, together with the engine sprocket and chain loop. Note that the clutch thrust washer (where fitted) must be fitted with its bronzed face outwards. The engine sprocket must be positioned so that the taper ground boss is closest to the crankshaft main bearing; the sprocket is a light drive fit on the crankshaft splines. Tap the clutch centre to lock it onto the gearbox mainshaft.

3 Fit the clutch inner drum over the splines of the clutch centre and replace the cup washer and self-locking nut that retains the assembly in position. Scotch the clutch inner drum and tighten the centre nut with a torque wrench to a setting of 50 lb ft. Earlier models have a slightly different locking arrangement embodying a tab washer for locking the centre nut after it has been tightened.

4 Replace the clutch plates in their original order (see Fig. 3.1a

30.4 Position push rod tubes correctly before lowering cylinder head

31.1 Locate push rods with cam followers

31.5 Don't omit spring clips under rocker box nuts

or 3.1b in Chapter 3). Fit the domed pressure plate, after the clutch pushrod has been inserted in the centre of the hollow mainshaft, and replace the thimbles, clutch springs and clutch adjusting nuts. Tighten the adjusting nuts evenly until at least one half of the thread has engaged with the stud projecting from the clutch inner drum.

5 It should be noted that the position of the crankshaft spacer may vary between different models in that it may be fitted inboard or outboard of the engine sprocket. If the position was not noted during removal, some experimentation may be necessary to establish which position is correct. Check that the engine and clutch sprocket faces are parallel, and that the inner face of the sprocket does not foul the crankcase when the primary drive is assembled. Replace the two forward-mounted studs which were removed to make the withdrawal of the engine sprocket easier and the sleeve nut in the chaincase rear, through which the leads from the generator stator pass. Assemble the rotor, spacer and sprocket as described above. Refit the tab washer and the centre retaining nut and tighten the latter with a torque wrench to a setting of 30 lb ft. Bend the tab washer to lock the nut. The timing marks on the rotor must face outward.

6 Slip the distance pieces over the stator coil retaining studs and position the stator coil assembly so that the lead connecting the coils is at the top. Thread the lead through the sleeve nut in the rear of the chaincase, until it emerges from the back. Replace the split plug that seals the sleeve nut orifice and the rubber cap over the sleeve nut extension. Check that there is no possibility of the lead fouling the primary chain. The lead must emerge from the outer portion of the stator coil assembly and be held in the clip on the underside.

7 Replace the stator plate retaining nuts and washers, then check that the rotor does not foul the stator plate assembly when the engine is turned over. There should be a minimum clearance of 0.008 in between each of the stator coil pole pieces and the rotor. Tighten to a torque wrench setting of 20 lb ft (2.765 kg m).

8 Refit the primary chaincase outer cover temporarily to prevent damage whilst the engine unit is lifted back into the frame. Replace the spark plugs to prevent dirt or other foreign matter from entering the engine during this same operation.

9 Locate the chain tensioner bolt and with a finger through the top filler plug hole, adjust the tensioner until the chain has just a small amount of up and down movement. Replace the tensioner plug.

10 If there is need to lock the engine during the final stages of assembly, the chaincase outer cover can be left off until the engine is replaced in the frame.

33 Adjusting the valve clearances

1 The valve clearances can now be set to save time when the engine is being fitted in the frame.

2 Should this operation be done with the engine in the frame, for service purposes, **the engine must be cold** or the correct setting will not be achieved.

3 The tappet adjusters are on the four rocker arms. The clearances are:

	Inlet	Exhaust
Tiger 100R — Daytona	0.002 in (0.05 mm)	0.004 in (0.10 mm)
T100S — Tiger 100	0.002 in (0.05 mm)	0.004 in (0.10 mm)
3TA — 5TA	0.010 in (0.25 mm)	0.010 in (0.25 mm)

4 On later models there are four small caps in the side of the rocker boxes in place of the rocker box caps. Remove all four.

5 Turn the engine over until one of the inlet valves is fully open, then with a 0.002 inch (0.05 mm) feeler gauge, adjust the valve which is fully closed. Slacken off the locknut and place the feeler gauge between the top of the valve and the tip of the adjuster. Screw in the adjuster until the feeler gauge is a good sliding fit, and tighten the locknut. Check, after the locknut has

31.6 Tighten down in the correct sequence to avoid distortion

32.2 Fit clutch centre complete with clutch chainwheel

32.4 Check clutch plates are assembled in correct order

been fastened, that the feeler gauge is still just a sliding fit.

6 Turn over the engine until the valve which you have just adjusted is fully open, and repeat the operation.

7 Exactly the same method is used for the exhaust valves; this time using a 0.004 inch (0.10 mm) feeler gauge.

8 Earlier machines have no provision for a feeler gauge so other methods of checking valve clearances are used. Rotate the engine as before so that one inlet valve is fully open. Adjust the closed valve by backing off the adjuster until the slightest perceptible amount of movement is felt, or a faint click can be heard. Lock up the adjuster and check.

9 Repeat this for the other inlet valve.

10 The exhaust valve adjuster should take up all the play and then be backed off 1/8 of a turn (half a flat). Lock up and recheck.

11 Repeat for the other exhaust valve.

12 On 3TA and earlier 5TA models the clearance for all four valves is 0.010 inch (0.25 mm). Adjustment is made as before except that the adjuster is backed off ¼ turn (1 flat). Repeat this for all four valves.

34 Replacing the engine in the frame

1 Replacing the engine in the frame is a two-man job. It is preferable to lift in from the left hand side of the machine.

2 Place the engine roughly in position, then locate the engine bolts and engine plates. Leave them loose until all the bolts are inserted. Do not forget the spacer on the long engine bolt underneath the crankcase.

3 Tighten all the engine bolts. Bolt on all the auxiliary equipment, not forgetting the head steady. Replace and tighten the carburettors, exhaust pipes, coils, and reconnect the colour-coded wires from the points and alternator; reconnect the oil pipes, check all drain plugs for tightness. Refit the rocker oil feed pipe, taking care not to overtighten the domed nuts.

4 Refill the oil tank. Refill the gearbox. Refill the chaincase with the correct grade of new oil.

35 Final adjustments

1 Refit the final drive chain (check for wear - if worn, it should be replaced). If the chain is in good condition, it should be cleaned and oiled.

2 Replace the petrol tank and connect the fuel pipes. Remember to replace the tank rubbers in their correct order.

3 Replace the footrests, gear change and kickstart levers.

4 Reconnect the clutch cable and adjust until a small amount of play is apparent at the handlebar lever.

5 Reconnect the tachometer and speedometer cables and any other extras.

6 Time the ignition - refer to Chapter 5 for the relevant information.

36 Starting and running the rebuilt engine

1 Make a final check to ensure that all oil pipe levels are correct and the pipes are tightly fitted.

2 Make sure that all the electrical connections are properly made.

3 Switch on the ignition and run the engine for a few minutes slowly.

4 Check, by looking in the top of the oil tank, for the engine oil return.

5 If no oil appears, stop the engine immediately and investigate. To see if the oil pump is functioning properly, slacken a few threads on the oil pressure switch or pressure release valve and the oil should appear. If it does, then retighten.

6 Should no oil still be showing from the return pipe, disconnect the oil return pipe and force oil down it, with the aid of a pressure oil can, whilst kicking the engine over. This clears any

32.5 Don't forget distance piece before fitting rotor

33.5 Method of adjusting valve clearances

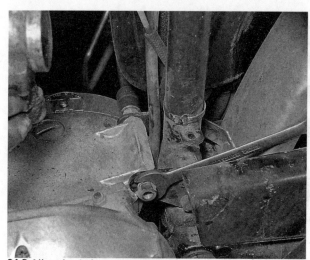

34.3 All engine bolts must be tightened fully

oil air lock and the oil should then pulse out of the tube. Reconnect the pipe to the oil tank and continue the check.

7   Check all controls for proper functioning, especially the clutch and brakes.

8   If new parts have been fitted the engine will require 'running in'. If a rebore has been performed, the running in period should be extended to at least 500 miles.

9   Do not tamper with the exhaust system to make the machine sound louder as a balance has been achieved by the factory for maximum performance with low noise level. If the exhaust is tampered with this can result in engine damage.

37 Engine modifications and tuning

1   The Triumph twin engine can be tuned to give even higher performance and yet retain a good standard of mechanical reliability. Many special parts for boosting engine performance are available both from the manufacturer and from a number of specialists who have wide experience of the Triumph marque. The parts available include high compression pistons, high lift camshafts and even cylinder heads with four valves per combustion chamber.

2   There are several publications, including a pamphlet available from the manufacturer, that provide detailed information about the ways in which a Triumph twin engine can be modified to give increased power output. It should be emphasised, however, that a certain amount of mechanical skill and experience is necessary if an engine is to be developed in this manner and still retain a good standard of mechanical reliability. Often it is preferable to entrust this type of work to an acknowledged specialist and therefore obtain the benefit of his experience.

34.3a Head steadies must also be tight

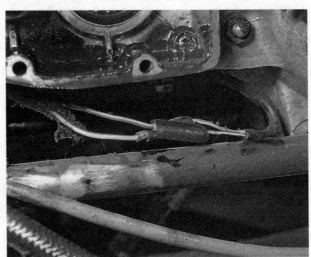

34.3b Colour-coding makes reconnection easy

35.3 Footrests are attached to rod through lower frame tubes

35.3a Pillion footrests serve also as means of silencer attachment

35.4 Slide cable stop holder cable before ....

35.4a .... engaging nipple with clutch actuating mechanism

35.4b Finally, fit slotted cable stop in end of holder

36.7 Clutch can be adjusted after removing plug in chaincase

38 Fault diagnosis

Symptom	Reason/s	Remedy
Engine will not turn over	Clutch slip	Check and adjust clutch.
	Mechanical damage	Check whether valves are operating correctly and dismantle if necessary.
Engine turns over but will not start	No spark at plugs	Remove plugs and check. Check whether battery is discharged.
	No fuel reaching engine	Check fuel system.
	Too much fuel reaching engine	Check fuel system. Remove plugs and turn engine over several times before replacing.
Engine fires but runs unevenly	Ignition and/or fuel system fault	Check systems as though engine will not start.
	Incorrect valve clearances	Check and reset.
	Burnt or sticking valves	Check for loss of compression.
	Blowing cylinder head gasket	See above.
Lack of power	Incorrect ignition timing	Check accuracy of setting.
	Valve timing not correct	Check timing mark alignment on timing pinions.
	Badly worn cylinder barrel and pistons	Fit new rings and pistons after rebore.
High oil consumption	Oil leaks from engine gear unit	Trace source of leak and rectify.
	Worn cylinder bores	See above.
	Worn valve guides	Replace guides.
Excessive mechanical noise	Failure of lubrication system	Stop engine and do not run until fault located and rectified.
	Incorrect valve clearances	Check and re-adjust.
	Worn cylinder barrel (piston slap)	Rebore and fit oversize pistons.
	Worn big end bearings (knock)	Fit new bearing shells.
	Worn main bearings (rumble)	Fit new journal bearings.

Chapter 2 Gearbox

Contents

Specifications

Model T100R — Daytona and T100C — Trophy 500

Ratios

Internal ratios (Std)

4th (Top)	1.00 : 1
3rd	1.22 : 1
2nd	1.61 : 1
1st (Bottom)	2.47 : 1

Overall ratios:

4th (Top)	5.70
3rd	6.95
2nd	9.18
1st (Bottom)	14.09
Engine rpm @ 10 mph in 4th (Top) gear	763
Gearbox sprocket teeth	18

Gear details

Mainshaft high gear:

Bore diameter (bush fitted)	0.7520 - 0.7530 in (18.9 - 18.93 mm)
Working clearance on shaft	0.0020 - 0.0035 in (0.0508 - 0.0889 mm)
Bush length	$2^{19/32}$ in (40.48 mm)

Layshaft low gear:

Bore diameter (bush fitted)	0.689 - 0.690 in (17.5 - 17.53 mm)
Working clearance on shaft	0.0015 - 0.003 in (0.038 - 0.0762 mm)

Gearbox shafts

Mainshaft:

Left end diameter	0.7495 - 0.7500 in (19.03 - 19.05 mm)
Right end diameter	0.6685 - 0.6689 in (16.98 - 16.99 mm)
Length	$9^{1/64}$ in (229 mm)
Length (before H.49833)	$8^{51/64}$ in (223.4 mm)

Layshaft:

Left end diameter	0.6845 - 0.6850 in (17.386 - 17.399 mm)
Right end diameter	0.6870 - 0.6875 in (17.45 - 17.5 mm)
Length	$5^{3/8}$ in (136.53 mm)

Camplate plunger spring:

Free length	2½ in (63.5 mm)
No of working coils	22
Spring rate	5 - 6 lb in (0.35 - 0.42 kg cm^2)

Model T100S — Tiger 100

Gear details
Layshaft low gear:
Bush protrusion ... ...   ¾ in (before H.57083) (9.525 mm)

Bearings
High gear bearing ... ...   30 x 62 x 16 mm Ball journal
Mainshaft bearing ... ...   17 x 47 x 14 mm Ball journal
Layshaft bearing (left) ... ...   11/16 x 7/8 x ¾ in Needle roller (17.46 x 22.23 x 19.05 mm)
Layshaft bearing (right) ... ...   5/8 x 13/16 x ¾ in Needle roller (15.87 x 20.64 x 19.05 mm)

Kickstart operating mechanism
Ratchet spring free length ... ...   ½ in (12.7 mm)

Gearchange mechanism
Plungers:
Outer diameter ... ...   0.3402 - 0.3412 in (8.64 - 8.66 mm)
Working clearance in bore ... ...   0.0015 - 0.0035 in (0.038 - 0.0889 mm)
Plunger springs:
No of working coils   16
Free length ... ...   1 1/16 in (26.99 mm)
Outer bush bore diameter ... ...   0.623 - 0.624 in (15.82 - 15.85 mm)
Clearance on shaft ... ...   0.001 - 0.003 in (0.0254 - 0.0762 mm)
Quadrant return springs:
No of working coils   18
Free length ... ...   1 7/8 in (47.63 mm)

Model T100T — Daytona Sports

Gearbox sprocket 18T

Model T90 — Tiger 90

Ratios
Internal ratios:
4th (Top) ... ...   1.00 : 1
3rd ... ...   1.22 : 1
2nd ... ...   1.61 : 1
1st (Bottom) ... ...   2.47 : 1
Overall ratios:
4th (Top) ... ...   6.03 : 1
3rd ... ...   7.36 : 1
2nd ... ...   9.71 : 1
1st (Bottom) ... ...   14.90 : 1
Engine rpm @ 10 mph in 4th gear   805
Gearbox sprocket teeth   17

Model 5TA — Speed Twin

Ratios
Internal ratios:
4th (Top) ... ...   1.00 : 1
3rd ... ...   1.22 : 1
2nd ... ...   1.61 : 1
1st (Bottom) ... ...   2.47 : 1
Overall ratios:
4th (Top) 5.40
3rd 6.59
2nd 8.69
1st (Bottom) 13.34
Engine rpm @ 10 mph in 4th (Top) gear   720
Gearbox sprocket teeth   20
Layshaft bushes:
Material ... ...   Bronze
Bore size, LH ... ...   0.6865 - 0.6885 in (17.44 - 17.487 mm)
Bore size, RH ... ...   0.690 - 0.689 in (17.53 - 17.56 mm)
Interference fit in casing, RH ... ...   0.0005 - 0.0015 in (0.0127 - 0.038 mm)
Interference fit in kickstart assembly, LH ... ...   0.0005 - 0.0015 in (0.0127 - 0.038 mm)

Model 3TA — Twenty One

Ratios

Internal ratios (close):		
4th (Top)	1.00 : 1
3rd	1.12 : 1
2nd	1.35 : 1
1st (Bottom)	1.99 : 1
Internal ratios (wide):		
4th (Top)	1.00 : 1
3rd	1.37 : 1
2nd	1.97 : 1
1st (Bottom)	3.18 : 1
Gearbox sprocket	17, 18, 19 and 20 teeth

Unless otherwise stated, T100R data applies to all models

1 General description

It is not necessary to strip the engine before the gearbox can be dismantled. The gearbox is easy to work on, and will normally require attention only after a very high mileage has been covered, or if the oil level has not been maintained.

To meet the demands of high performance ie road racing, there are available close ratio gear clusters which, with little or no modification, will replace the standard gears. For details of such a change over, consult a dealer who has the appropriate Triumph replacement parts lists.

2 Dismantling the gearbox - removing the outer cover

1   Clean the gearbox outer case and have the appropriate tools to hand.
2   Remove the right hand exhaust system as a complete unit.
3   Drain off the gearbox oil into a drip tray. Slacken the clutch handlebar lever adjustment; unscrew the barrel which goes into the top of the gearbox, take off the rubber cover. Pull out the barrel and slip the cable nipple out of the slot in the clutch operating mechanism.
4   Engage top gear - this helps lock the nuts which have to be slackened at a later stage.
5   Unscrew the two nuts and four recessed screws from the periphery of the outer cover. Remove the kickstart lever.
6   Grasp the footchange lever with the left hand and holding a hide mallet or soft faced object in your right hand, tap the outer cover until it comes away from the machine.

3 Dismantling the gearbox - removing the inner cover and the gear clusters

1   To remove the gearbox inner cover it is necessary to detach the primary side chaincase outer cover and clutch as described in the engine strip-out, Section 9, paragraphs 1 to 13.
2   It is best to renew the tab washer, which will have to be bent back to allow the gearbox mainshaft nut to be undone. Engage between 2nd and 3rd gear before taking off the outer case.
3   Remove the gearbox inner cover by slackening and taking out the two remaining screws. Tap the end of the mainshaft with a hide mallet to drift out the cover and gear clusters.

4 Removing the camplate and gear clusters from the gearbox main cover

1   The gearbox (camplate) can be removed by withdrawing the split pin which passes through the camplate pivot.
2   The gear clusters can now be examined for wear of the dogs and ramps, also for worn gear teeth. It is not necessary to strip the cluster if only a check is being carried out.
3   Lever off the kickstart plate spring, and distance piece, which will allow the kickstart spindle to be removed.
4   It is also advisable to extract the cam plunger situated in the back of the gearbox, at the top. Keep it in a safe place, and do not forget to replace it **before** reassembly.
5   Take out the camplate after disengaging it from the selector forks. Then disengage the selector forks from the gear cluster.
6   Withdraw the layshaft from the inner case leaving the mainshaft in position.
7   Drive the mainshaft out with a hide mallet.

5 Examination of the gearbox components - general

1   The end cover should be inspected for signs of cracking and distortion or damage to bearing and gasket surfaces. If the castings are cracked, have them checked by a specialist welder; he will be able to tell you whether it is cheaper to weld it or buy a new or secondhand replacement.
2   If there is any doubt about any component, it is always better to replace it, because a breakdown far from home can be both inconvenient and costly.

6 Dismantling the gearbox - removing and replacing the mainshaft and layshaft bearings

1   To allow the gearbox main drive top gear to be removed, it is necessary to remove the large round plate situated behind the clutch, and fold back the tab washer. Remove the large nut holding the gearbox final drive sprocket and pull the sprocket off the splines.
2   The top gear and combined sleeve gear can now be driven into the gearbox with a hide hammer.
3   The sleeve gear bearing is removed by levering out the oil seal (always fit a new replacement) and removing the circlip. Heat the gearbox shell around the bearing and drive out the old bearing, then drive in the new replacement before the shell cools down. Replace the circlip and fit the new oil seal.
4   The other mainshaft bearing is in the inner gearbox cover; remove the circlip, heat the cover and drive out the bearing, then refit the new replacement before the case cools down. Replace the circlip.
5   To remove the layshaft needle roller bearings from the gearbox shell, heat the shell and tap it out, then fit the new bearing before the shell cools down.
6   The other layshaft bearing is housed in the kickstart gear assembly. Heat the assembly and with a short, sharp bang, shock out the old bearing out of position before the new replacement is drifted in.

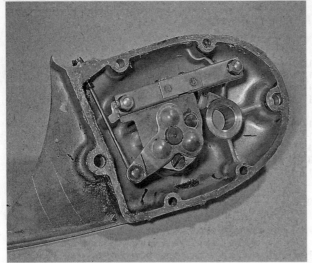

2.6 Clutch operating mechanism is within outer cover

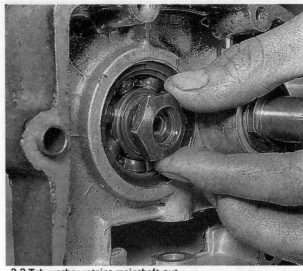

3.2 Tab washer retains mainshaft nut

3.3. Gear clusters will withdraw with inner cover

3.3a The gear clusters, complete with cam plate and selectors

4.1 A split pin retains camplate spindle

4.4 Camplate plunger remains in gearbox shell

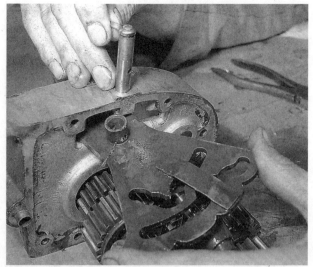

4.5 Withdraw spindle to free camplate

4.5a Selectors will lift off gear pinions

4.6 Layshaft withdraws from centre of kickstarter pinion

4.7 Tap mainshaft out of main bearing centre

6.1 Circular cover gives access to final drive sprocket

6.1a Nut retaining final drive sprocket

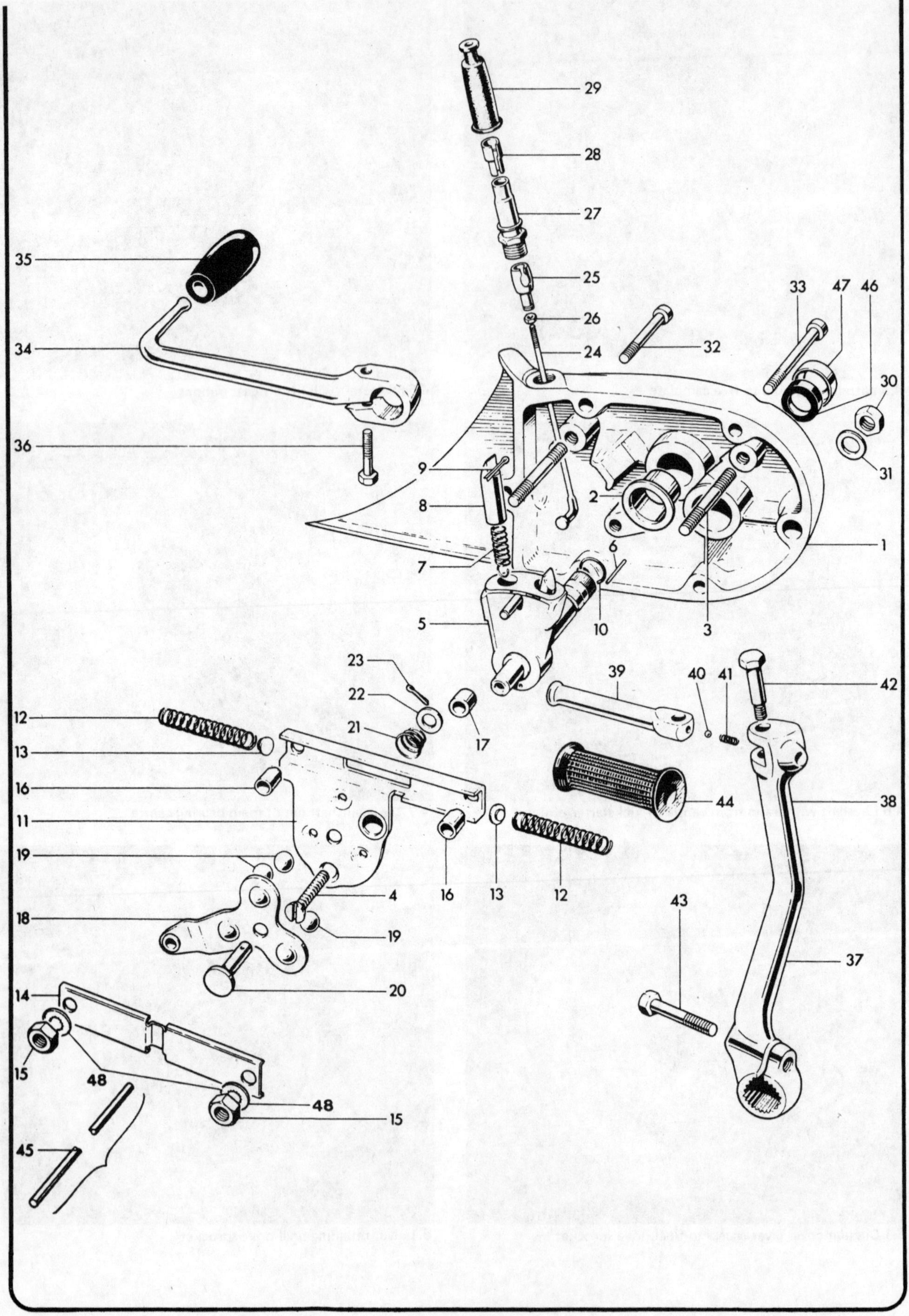
29
28
27
35
25
33
47
46
26
32
34
24
30
36
31
9
2
8
1
7
6
5
10
3
23
39
40
41
42
22
12
21
17
13
16
44
38
11
19
16
13
12
4
43
18
19
37
20
14
15
48
48
15
45

Fig. 2.1. Gearbox outer cover and levers

1	Gearbox outer cover 1 off
2	Gearchange spindle bush 1 off
3	Stud 2 off
4	Screw 1 off
5	Gearchange quadrant 1 off
6	Spring retaining pin 2 off
7	Plunger spring 2 off
8	Selector plunger 2 off
9	Split pin 2 off
10	Sealing ring 1 off
11	Return spring housing 1 off
12	Return spring 2 off
13	Thrust button 2 off
14	Cover 1 off
15	Nut 3 off
16	Distance piece 2 off
17	Distance piece 1 off
18	Clutch lever 1 off
19	Ball, 3/8 in. diam. 3 off
20	Shaft 1 off
21	Spring 1 off
22	Plain washer 1 off
23	Split pin 1 off
24	Spoke 1 off

25	Connector 1 off
26	Nut 1 off
27	Abutment 1 off
28	Adaptor, short
28	Adaptor, long (for cable with short outer cover)
29	Rubber cover 1 off
30	Nut 2 off
31	Plain washer 2 off
32	Screw 2 off
33	Screw 2 off
34	Gearchange lever 1 off
35	Pedal rubber 1 off
36	Bolt 1 off
37	Kickstarter lever complete 1 off
38	Kickstarter crank 1 off
39	Kickstarter pedal 1 off
40	Steel ball (¼ in. diam.) 1 off
41	Index spring 1 off
42	Pivot bolt 1 off
43	Clamp bolt 1 off
44	Pedal rubber 1 off
45	Clutch operating rod 1 off
46	Oil seal 1 off
47	Oil seal cover
48	Serrated washer 2 off

7 Kickstarter mechanism - examination and renovation

1   The kickstarter is a much used component and malfunction will result in many problems.

2   To extract the kickstart shaft, lever off the spring and plate, as in Section 4, paragraph 3, of this Chapter.

3   Wear of the kickstarter pawl will result in the kickstarter slipping. If the pawl is worn, replace it together with the spring and plunger (all these parts are reasonably cheap but they have a hard life).

4   The teeth in which the pawl engages should also be examined closely for signs of wear. If rounded, the gear pinion must be replaced. (This part is reasonably expensive because it doubles as the layshaft low gear.)

5   When reassembling, it is recommended that the kickstart assembly is inserted in the layshaft low gear and the inner cover replaced afterwards.

6   When renewing the kickstart return spring, care must be taken to tension the spring properly. The kickstart lever's own weight will otherwise allow it to drop back when accelerating.

8 Gearchange mechanism and clutch operating mechanism - examination and renovation

1   There is usually no need to strip the gear change mechanism unless it is faulty, ie bad selection, not selecting, etc.

2   The same applies to the clutch operating mechanism which gives trouble only if a breakage occurs or if there is a clicking noise from the mechanism when operated. This latter defect signifies the three ball bearings or the cups in which they seat, are worn out. Renewal of the worn parts is the only remedy.

3   To strip the gear change mechanism, it is first necessary to take off the clutch operating mechanism. Unscrew the two nuts inside the outer gearbox cover and withdraw the cover plate foot change return spring, complete with the two thrust buttons and distance pieces.

4   Unscrew the countersunk screw which retains the clutch operating mechanism and withdraw the unit.

5   Loosen the retaining bolt which holds the gearchange lever and tap the end of the splines lightly thus removing the gear change assembly.

6   Inspect the gear change plungers for wear and check whether they are still a good fit in the block. Renew the springs if they are broken or weak. Check the tips and also the slots in the camplate; if any part is badly worn, it must be renewed. Inspect the gear change return springs for corrosion, breakage or weakness. Replace them if they are not in good condition.

7   Check the bush in which the quadrant operates. Renew if it is oval or scored badly. To renew, heat the cover, tap out the old bush and tap in the new one before the case cools down.

8   When assembling the gear change unit into the cover, renew the O ring in the cover and oil lightly to aid reassembly. Retighten all nuts and screws.

9 Gearbox components - examination and renovation

1   Examine the layshaft for signs of wear and fatigue. If in doubt, check the diameters and dimensions with the Specifications data. Any discolouring or blueing of a shaft means it has been running excessively hot, probably through lack of oil.

2   Check all the splines for signs of wear and for burrs. If they are only slightly marked, they can be removed by oil stoning. If they are more badly worn then the parts concerned must be replaced.

3   If any bushes are worn, or needle rollers are damaged, renewal is essential.

4   The ball journal bearings, when washed out with paraffin, must run smoothly. Should any roughness be apparent, the bearings must be replaced.

5   If there is any tendency for the gears to bind on their shafts,

6.2 Sleeve gear bearing after removal of sleeve gear

6.3 Sleeve gear bearing is retained by a circlip

6.3a Sleeve gear bearing oil seal in position

treatment with an oilstone is necessary. If they slide too freely and appear slack, then renewal is necessary due to the high loads experienced by the gearbox.

6 If any bearings have been turning in their housing, Loctite will suffice if the amount of play is only minimal. In cases where play is excessive, then only new or good secondhand cases will provide a cure.

10 Gearbox reassembly - general

1 All surfaces must be clean.
2 All components must be washed clean with paraffin, then oiled liberally, especially bearings and the splines on which the gears slide.
3 Check all threads are clean and in good condition, and that all the locating pegs or dowels are in place.

11 Replacing the sleeve gear, final drive sprocket and selector plunger

1 The sleeve gear bearings and oil seal should be in position. Oil the centre of the oil seal with clean oil and tap the sleeve gear through until it is right home.
2 Oil the tapered side of the sprocket and push it into the oil seal.
3 Replace the tab washer and screw on the retaining nut, finger tight.
4 Connect the rear chain and using the rear brake to prevent the sprocket from turning, tighten the nut very hard.
5 Finally, flatten the tab washer against the flat on the nut, to stop the nut from slackening off.
6 Replace the spring and plunger in the hole above the sleeve gear using grease to hold it in position.

12 Replacing the gear clusters in the gearbox inner cover

1 With the gearbox inner cover assembled with all its bearings in the correct place, replace the layshaft thrust washer over the needle roller cage and hold it in position with a smear of grease.
2 Lubricate the mainshaft and layshaft captive gears, then assemble the mainshaft (as shown in the diagram) into the inner cover.
3 Refit the plunger spring, plunger and kickstart pawl to the kickstart spindle, and insert the complete assembly into the inner cover, sliding the layshaft assembly into the kickstarter bearing. **Do not forget to fit the mainshaft assembly spacer**, located between the bearing and mainshaft.
4 The selector forks, although looking similar, differ slightly. Prior to the final assembly, check, by manually moving the selector plate, whether the selectors move to their full extremity in each groove. If not, pull out the spindle and change round the selectors.
5 Position the selector forks correctly and slide the spindle into the inner cover so that it locates in its housing.
6 Assemble the cam plate into the inner cover and locate the slots over the selector rollers.
7 Fit the camplate spindle and insert the split pin. Oil liberally. The inner cover is now ready for reassembly with the gearbox shell.

13 Fitting the gearbox inner cover and the kickstarter spring

1 Apply a layer of gasket cement to the jointing faces of the inner cover and the gearbox shell. Insert the complete assembly, taking care not to displace the thrust washer at the back of the gearbox or the camplate plunger. Locate with the dowel pin and tap the assembly home. If heavy resistance is felt, it may be necessary to rock the gear cover to aid location. Insert and tighten the fixing screws.

7.2 Kickstarter pinion will extract from inner cover

7.4 Pawls with rounded edges must be replaced

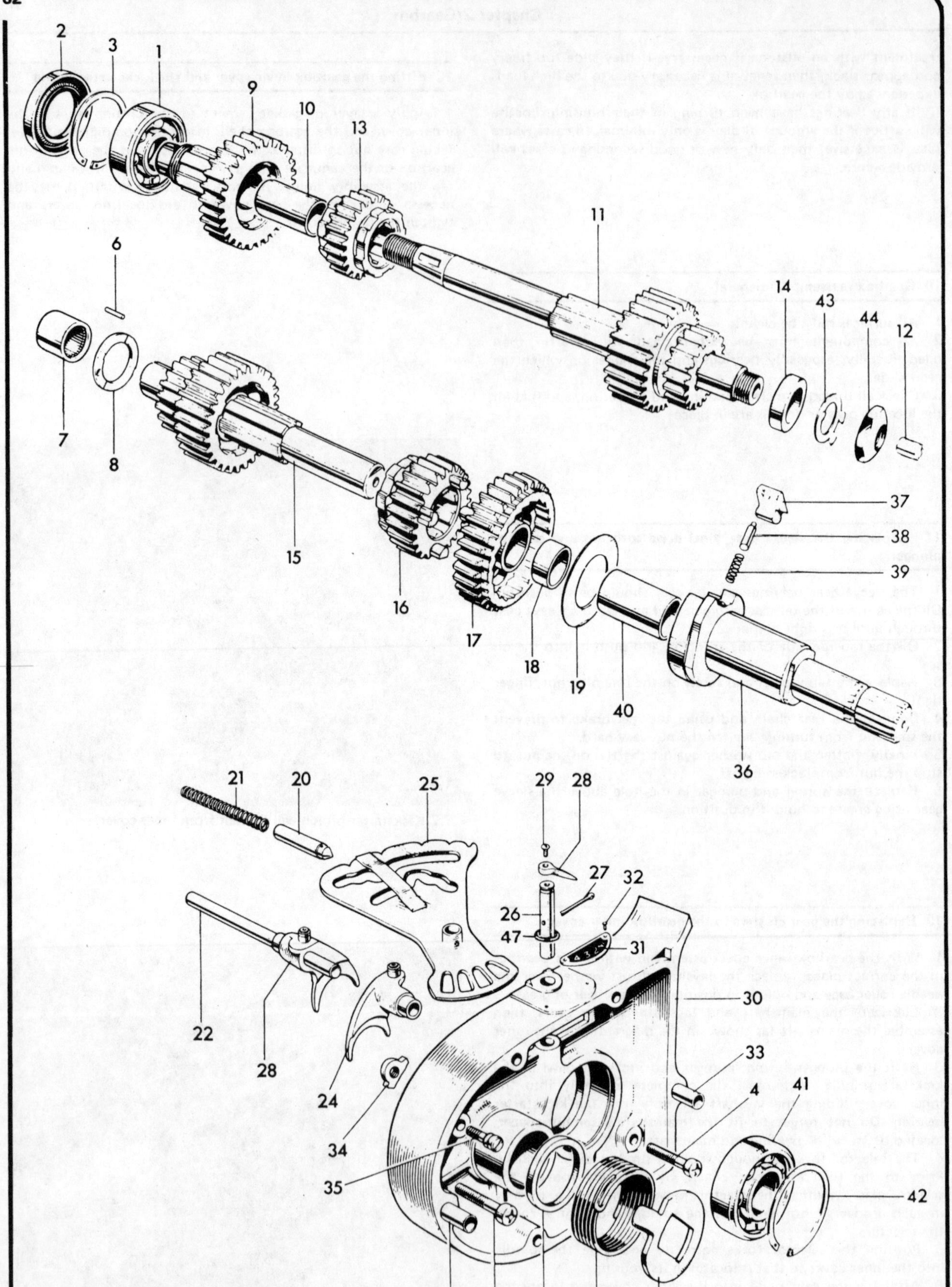
2
3
1
9
10
13
11
6
14
43
44
12
7
8
15
16
17
18
19
40
36
37
38
39
21
20
25
29
28
27
32
26
47
31
22
28
30
24
33
4
41
34
35
42
33
4
45
46
5

Fig. 2.2. Gearbox shafts, bearings and inner cover

1	High gear bearing (ball journal) 1 off	25	Gear selector camplate 1 off
2	Oil seal 1 off	26	Camplate spindle 1 off
3	Circlip 1 off	27	Split pin 1 off
4	Screw 2 off	28	Gear indicator pointer 1 off
5	Return spring plate 1 off	29	Screw 1 off
6	Peg 1 off	30	Inner cover 1 off
7	Layshaft bearing 1 off	31	Gear indicator plate 1 off
8	Thrust washer 1 off	32	Indicator plate rivet 2 off
9	Mainshaft high gear (22T) 1 off	33	Hollow dowel 2 off
10	High gear bush 1 off	34	Kickstarter stop plate 1 off
11	Mainshaft c/w third and low gears (16/24T)	35	Stop plate and anchor screw 1 off
12	Clutch rod bush 1 off	36	Kickstarter spindle 1 off
13	Mainshaft second gear (21T) 1 off	37	Kickstarter pawl 1 off
14	Mainshaft distance piece 1 off	38	Plunger 1 off
15	Layshaft c/w second and high gears (15/23T) 1 off	39	Plunger spring 1 off
16	Layshaft third gear (20T) 1 off	40	Bearing 1 off
17	Layshaft low gear (27T) 1 off	41	Mainshaft bearing (ball journal) 1 off
18	Low gear bush 1 off	42	Circlip 1 off
19	Pawl retaining disc 1 off	43	Tab washer 1 off
20	Camplate index plunger 1 off	44	Mainshaft nut 1 off
21	Index plunger spring 1 off	45	Distance piece 1 off
22	Selector fork spindle 1 off	46	Kickstarter return spring 1 off
23	Mainshaft selector fork 1 off	47	"O" ring
24	Layshaft selector fork 1 off		

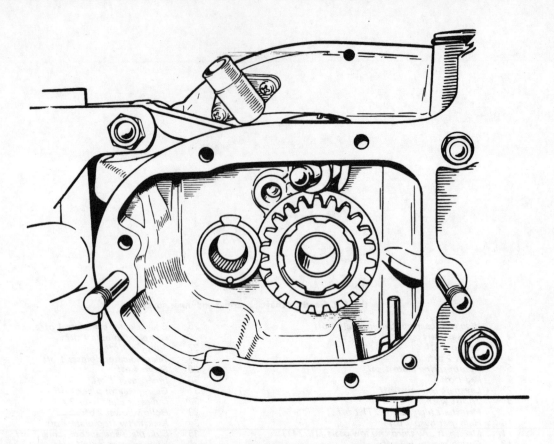

Fig. 2.3. Location of camplate plunger, layshaft bush and high (sleeve) gear

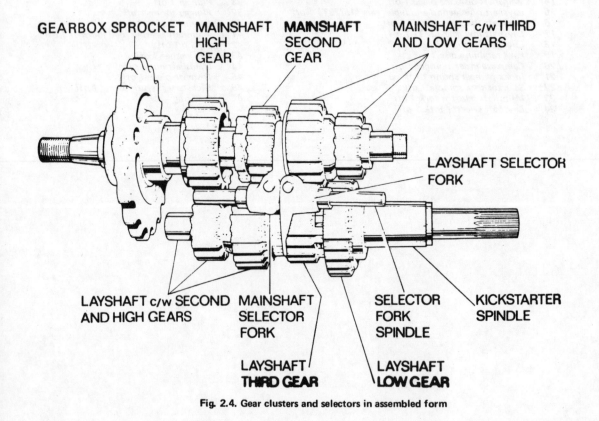

Fig. 2.4. Gear clusters and selectors in assembled form

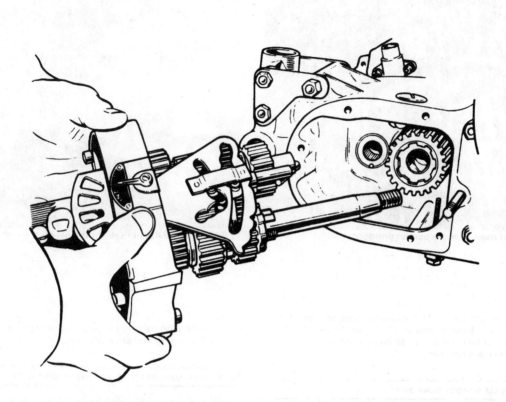

Fig. 2.5. Reassembling the gearbox inner cover assembly into the gearbox shell

9.5 A good example of badly worn dogs

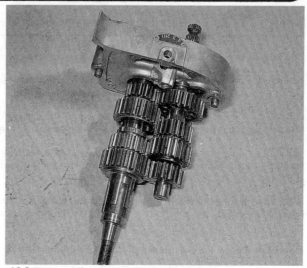

12.3 Reassembling the gear cluster in the inner cover

12.5 Selectors must be in the correct positions

12.7 Use a new split pin to retain the camplate spindle

2 Replace the spacer over the kickstarter shaft, the spring and plate. Use a screwdriver to tension the spring. Locate the plate and hook around location screw. Be very careful not to disturb when replacing the outer cover.

14 Replacing the gearbox outer cover

1 Place the machine in neutral and apply a layer of gasket cement to the cover face. Oil the protruding shaft and fit the outer cover, tapping it right home.
2 Replace the screws and nuts, and tighten fully.
3 Replace the kickstart lever and gearchange lever.
4 Refit the footrest, exhaust pipe and other auxiliary equip-

ment.
5 Refill the gearbox with the correct quantity of oil.
6 Check that all gears can be selected.

15 Reassembling the clutch and engine sprocket housing

1 Using a new gasket, replace the round final drive sprocket cover. Tighten the six screws around the periphery.
2 Replace the clutch, engine sprocket and primary chain as a complete unit.
3 Fit the alternator stator and the chaincase outer cover. Adjust the primary chain and replenish the oil in the chaincase.

16 Fault diagnosis

Symptom	Reason/s	Remedy
Difficulty in engaging gears	Gears not indexed correctly	Check timing sequence of inner end cover (will occur only after rebuild).
	Worn or bent gear selector forks	Examine and renew if necessary.
	Worn camplate	Examine and renew as necessary.
	Low oil content	Check gearbox oil level and replenish.
Machine jumps out of gear	Mechanism not selecting positively	Check for sticking camplate plunger or gear change plungers.
	Sliding gear pinions binding on shafts	Strip gearbox and ease any high spots.
	Worn or badly rounded internal teeth in pinions	Replace all defective pinions.
Kickstarter does not return when engine is started or turned over	Broken kickstarter return spring	Remove outer end cover and replace spring.
	Kickstarter ratchet jamming	Remove end cover and renew all damaged parts.
Kickstarter slips on full engine load	Worn kickstarter ratchet	Remove end cover and renew all damaged parts.
Gear change lever fails to return to normal position	Broken or compressed return springs	Remove end cover and renew return springs.

Chapter 3 Clutch

Contents

Specifications

Model T100R — Daytona and T100C — Trophy 500

Clutch details

Type	Multiplate with integral shock absorber
No of plates:	
Driving (bonded)	6
Driven (plain)	6
Pressure springs:	
Number	3
Free length	1^{31}/32 in (50 mm)
No of working coils	9½
Spring rate	58½ lb in (4.1 kg cm^2)
Approximate fitted load	42 lb in (2.95 kg cm^2)
Bearing rollers:	
Number	20
Diameter	0.2495 - 0.2500 in (6.34 - 6.35 mm)
Length	0.231 - 0.236 in (5.87 - 5.98 mm)
Clutch hub bearing diameter	1.3733 - 1.2742 in (9.48 - 9.5 mm)
Clutch sprocket bore diameter	1.0745 - 1.0755 in (25.4189 - 25.942 mm)
Thrust washer thickness	0.052 - 0.054 in (0.002 - 0.0021 mm)
Engine sprocket teeth	26
Clutch sprocket teeth	58
Chain details	Duplex endless ^{3}/8 in pitch x 78 links

Clutch operating mechanism

Conical spring:	
No of working coils	2
Free length	1^3/32 in (10.32 mm)
Diameter of balls	^{3}/8 in dia (9.53 mm)
Clutch operating rod:	
Diameter of rod	^{3}/16 in dia (4.76 mm)
Length of rod	9.562 - 9.567 in (242.87 - 243.0 mm)

Model 3TA — Twenty One

Clutch details

No of plates:	
Driving (bonded)	5
Driven (plain)	5
Length of clutch operating rod	9.432 - 9.442 in (239.57 - 239.83 mm)

Unless otherwise stated, T100R data applies to all models

1 General description

The clutch is of the multi-plate type, designed to operate in oil. A synthetic friction material is used to line the inserted plates and a rubber cushion shock absorber is incorporated inside the centre drum of the clutch to even out transmission surges, particularly at low speeds. Drive from the engine sprocket is transmitted by a duplex roller chain. Because the gearbox is in unit with the engine, it is not possible to vary the centres between the engine mainshaft and the gearbox mainshaft. In consequence, a chain tensioner is incorporated so that the chain can be adjusted at regular intervals to take up wear.

2 Adjusting the clutch

1 The clutch can be adjusted by means of the handlebar lever cable adjuster, the cable adjuster in the top of the gearbox end cover, the pushrod adjuster in the centre of the clutch pressure plate and by varying the tension of the three clutch springs. In the latter case, the chaincase cover must be removed before adjustment can be effected.

2 To adjust the clutch, slacken off the handlebar lever adjuster so that there is an excess of free play in the lever, then unscrew the circular threaded plug in the outer face of the primary chaincase so that access can be gained to the adjuster in the centre of the pressure plate. Slacken the locknut and screw the adjuster inwards until the pressure plate just commences to lift. Back off the adjuster one complete turn, tighten the locknut and replace the threaded plug in the chaincase.

3 Adjust either the handlebar lever adjuster or the adjuster in the top of the gearbox end cover until there is approximately 1/8 in free play in the cable. This will ensure that there is no permanent loading on the clutch pushrod. Clutch adjustment should now be correct.

4 If the clutch still drags and the adjustment procedure described has produced no improvement, it will be necessary to remove the primary chaincase cover completely, in order to gain access to the three clutch spring adjusters in the pressure plate. Before slackening the adjuster nuts, check that the drag is not caused by uneven tensioning of the pressure plate. This check is made by using the handlebar lever to slip the clutch and turning the clutch by depressing the kickstarter. Uneven tension will immediately be obvious by the characteristic 'wobble' of the pressure plate. The wobble can be corrected by retensioning the clutch springs individually until the pressure is even.

5 If the clutch still drags, the adjusters should be slackened off an equal amount at a time before a recheck is made. Do not slacken them off too much, or the clutch operation will become very light with the possibility of clutch slip when the engine is running.

6 Drag is often caused by wear of the clutch outer drum. The projecting tongues of the inserted plates will wear notches in the grooves of the drum, which will eventually trap the inserted plates as the clutch is withdrawn and prevent them from separating fully. The accompanying photograph is an excellent example of such extended wear. In a case such as this, renewal of both the clutch chainwheel and the inserted clutch plates is necessary. If wear is detected in the early stages, it is possible to redress the grooves with a file until they are square once again, and to remove the burrs from the edges of the clutch plate tongues.

7 Heavy clutch operation is sometimes attributable to a cable badly in need of lubrication, or one in which the outer covering has become badly compressed through being trapped. Even a sharp bend in the cable will stiffen up the operation.

8 Clutch slip will occur when the clutch linings reach their limit of wear. If the reduction in lining thickness exceeds 0.030 in slip will occur and the inserted plates must be renewed. This necessitates dismantling the clutch, as described in Chapter 1, Section 9. There is no necessity to dismantle any part of the generator or primary transmission assembly because the clutch plates can be withdrawn after the clutch pressure plate is removed.

9 It cannot be overstressed that a great many clutch problems are caused by failure to maintain the oil content of the chaincase at the correct level or by the use of a heavier grade of oil than that recommended by the manufacturers. It is also important that the chaincase oil is changed at regular intervals, to offset the effects of condensation.

3 Examining the clutch plates and springs

1 When the clutch is dismantled for replacement of the clutch plates or during the course of a complete overhaul, this is an opportune time to examine all clutch components for signs of wear or damage.

2 As mentioned previously, the inserted clutch plates will have

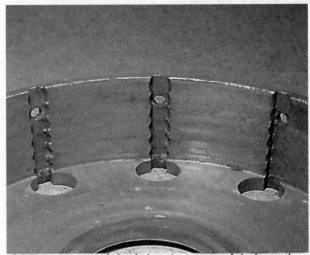

2.6 Excessive wear of clutch drum by tongues of the inserted plates

2.8. Check thickness of the inserted plates when the clutch is stripped

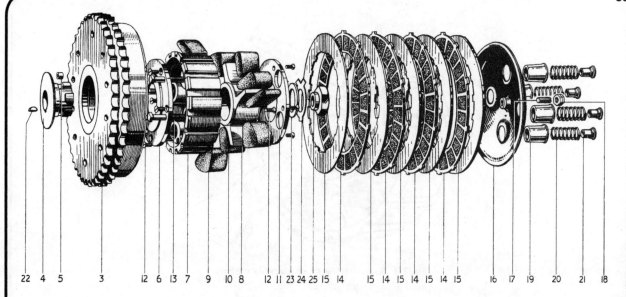

Fig. 3.1a Clutch assembly – pre 1963 models

3 Chainwheel and clutch
  outer drum
4 Clutch centre
5 Roller – 20 off
6 Inner plate
7 Clutch inner drum

8 Shock absorber spider
9 Drive rubber – 4 off
10 Rebound rubber – 4 off
11 Outer cover
12 Countersunk screw – 8 off
13 Screwed pin – 4 off

14 Inserted plate
15 Plain plate
16 Pressure plate
17 Adjuster pin
18 Locknut
19 Spring thimble – 4 off

20 Clutch spring – 4 off
21 Nut (brass) – 4 off
22 Key
23 Cupped washer
24 Tab washer
25 Clutch nut

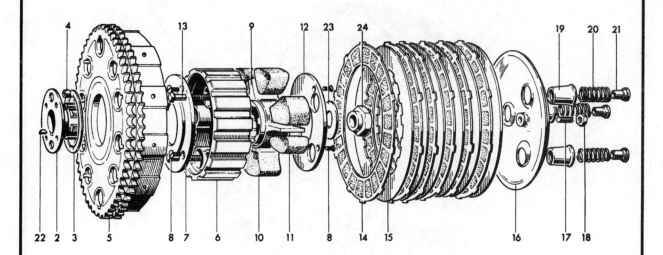

Fig. 3.1b Clutch assembly – 1963 on models

2 Clutch centre
3 Thrust washer
4 Roller - 20 off
5 Chain wheel and clutch
  outer drum (58 teeth
  duplex)
6 Clutch inner drum
7 Inner plate

8 Countersunk screws
  - 6 off
9 Shock absorber spider
10 Drive rubber (large) - 3 off
11 Rebound rubber (small)
  - 3 off
12 Outer cover
13 Screwed pin - 3 off

14 Inserted plate
  - 6 off
15 Plain plate - 6 off
16 Pressure plate
17 Adjuster pin
18 Locknut
19 Clutch spring thimble
  - 3 off

20 Clutch spring - 3 off
21 Clutch spring nuts
  (brass) - 3 off
22 Key
23 Cupped washer
24 Clutch nut

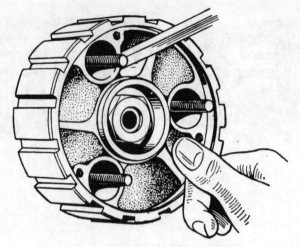

Fig. 3.2. Replacing the shock absorber rubbers

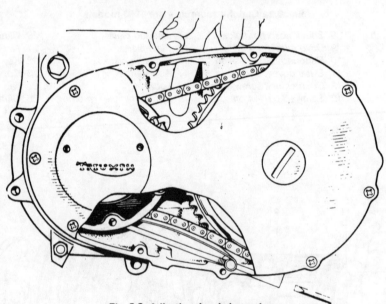

Fig. 3.3. Adjusting the chain tensioner

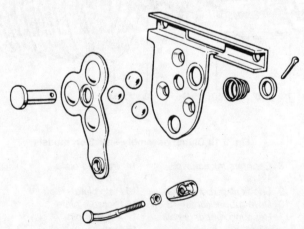

Fig. 3.4. Exploded view of clutch operating mechanism

to be renewed when the reduction in thickness of the linings reached 0.030 in. It is important to check the condition of the tongues at the edge of each plate which engage with the grooves in the clutch chainwheel. Even if the clutch linings have not approached the wear limit, it is advisable to renew the inserted plates if the width of the tongues is reduced as the result of wear.

3 Both plain and inserted plates should be perfectly flat. Check by laying them on a sheet of plate glass. Discoloration will occur if the plates have overheated; the surface finish of the plates is not too important provided it is smooth and the plates are not buckled.

4 The clutch springs will compress during service and take a permanent set. If any spring has reduced in length by more than 0.2 in, all three springs should be renewed - never renew one spring on its own. See Specifications Section for the original free length measurement.

5 Check the hardened end of the clutch adjuster in the centre of the clutch pressure plate to ensure it has not softened from overheating or that the hardened surface is not chipped, cracked or worn away. This also applies to the ends of the clutch pushrod. Mysterious shortening of the pushrod, necessitating frequent clutch adjustment, can usually be attributed to incorrect adjustment of the clutch, such that the absence of free play in the clutch cable places a permanent load on the pushrod. This in turn causes the ends of the pushrod to overheat and soften, thus greatly accelerating the rate of wear.

6 The teeth of the clutch chainwheel should be examined since chipped, hooked or broken teeth will lead to very rapid chain wear. It is not possible to reclaim a worn chainwheel; the whole assembly must be renewed.

3.4 Check clutch springs for wear by measuring length

4 Examining the clutch shock absorber assembly

1 The shock absorber assembly is contained within the clutch inner drum and can be examined by unscrewing the three countersunk screws in the front cover plate. Remove the cover plate using a small screwdriver as a lever.

2 The shock absorber rubbers can be prised out of position commencing with the smaller rebound rubbers, to make the task easier. Avoid damage to the rubbers which may disintegrate in service if they are punctured or cracked. The centre 'spider' will be left in position and need not be disturbed unless it is cracked or broken. It is held in position by the nut that retains the clutch assembly to the gearbox mainshaft.

3 Reassemble the shock absorber assembly by reversing the dismantling procedure. Insert the large drive rubbers first and follow up with the smaller rebound rubbers. Fitting will be made easier if the smaller rubbers are smeared with household liquid detergent so that they can be slid into position. It is advisable to keep the rubbers free from oil, even though they are made of a synthetic material.

4 Before the cover plate screws are replaced, smear their threads with a sealant such as Loctite and tighten them fully. It is permissible to caulk the heads in position with a centre punch.

3.5. Hardened ends of push rod must be in good condition

5 Adjusting the primary chain

1 The primary chain is of the duplex type and is endless because the engine and gearbox are built in unit with fixed centres. Provision for chain adjustment is made by incorporating a Weller type chain tensioner in the chaincase which is adjustable from a tunnel cast longitudinally in the base of the chaincase and sealed off by the chaincase drain plug.

2 To adjust the chain tensioners, first remove the filler cap from the top of the chaincase, to the rear of the cylinder barrel. Remove the chaincase drain plug, after placing a container below, to catch the oil. On some models it may be necessary to slacken the left hand footrest in order to gain better access.

3 Insert a screwdriver into the chaincase tunnel and turn it clockwise to increase the chain tension. Tension is correct when

3.6 Broken or chipped teeth cause rapid chain wear

there is ½ inch free movement in the centre of the top chain run. Check with the sprockets in several different positions because a chain seldom wears evenly. The ½ inch play should be at the tightest point in the chain run.

4   Replace the chaincase drain plug, refill the chaincase with the correct amount of oil and replace the filler cap. Make sure the left hand footrest is tightened again if it has been slackened off.

5   If the chain tensioner reaches the end of its adjustment before the correct chain tension is achieved, the chain is due for renewal. It would be advisable to renew both the engine sprocket and the clutch chainwheel at the same time whilst the primary transmission is dismantled completely. If old and new parts are run together, there is every possibility that the rate of wear will be accelerated.

6   The chain tensioner itself is unlikely to require attention unless the rubber facing becomes detached from the tensioner slipper. If this occurs, the tensioner should be renewed at the earliest opportunity. Damage of this nature is most likely to be caused if the chaincase is allowed to run low in oil content.

6 Fault diagnosis

Symptom	Reason/s	Remedy
Engine speed increases but machine does not respond	Clutch slip	Check clutch adjustment. If correct, suspect worn linings and/or weak springs.
Difficulty in engaging gears. Gear changes jerky and machine creeps forward, even when clutch is withdrawn fully	Clutch drag Clutch plates worn and/or clutch drums Clutch assembly loose on mainshaft	Check adjustment for too much play. Check for burrs on clutch plate tongues and indentations in clutch drum grooves. Check tightness of retaining nut. If loose, fit new tab washer and retighten.
Operating action stiff	Damaged, trapped or frayed control cable Cable bends too acute Pushrod bent Spring adjusters too tight	Check cable and replace if necessary. Re-route cable to avoid sharp bends. Replace. Slacken adjusters and check clutch does not slip.
Clutch needs frequent adjustment	Rapid wear of pushrod	Leave slack in cable to prevent continual load on pushrod. Renew rod because overheating has softened ends.
Harsh transmission	Worn chain and/or sprockets	Replace.
Transmission surges at low speeds	Worn or damaged shock absorber rubbers	Dismantle clutch shock absorber and renew rubbers.

Chapter 4 Fuel system and lubrication

Contents

Specifications

T100R

Carburettor	USA	Home and General Exp.	Before H 57083	After H 57083
	Concentric	Concentric	Monobloc	Concentric
Amal type	626/36-38	626/54	376/324 and 325	626/9 and 10
Main jet size	170	150	200	140
Pilot jet size	–	–	25	–
Needle jet size	0.106	0.106	0.106	0.106
Needle type	–	–	C	Two rings above needle clip grooves *
Needle position	1	1	3	2
Throttle valve type	3½	3	376/3½	3
Carburettor nominal bore size	26 mm	26 mm	1^{1}/16 in	26 mm
Air cleaner type	–	Paper	Coarse felt	Coarse cloth

* An alternative needle is available for use with alcohol, marked Y

T100C

Carburettor

Main jet	170
Needle jet	0.106
Throttle valve	4
Needle position	2

T100R and T100C

Oil pump

Body material	Brass
Bore diameter:	
Feed	0.3748 - 0.3753 in (9.529 - 9.532 mm)
Scavenge	0.4372 - 0.4377 in (11.105 - 11.118 mm)
Scavenge (before H 49833)	0.4877 - 0.4872 in (12.375 - 12.388 mm)
Plunger diameter:	
Feed	0.3744 - 0.3747 in (9.5098 - 9.5174 mm)
Scavenge	0.4369 - 0.4372 in (11.097 - 11.105 mm)
Scavenge (before H 49833)	0.4872 - 0.4869 in (12.375 - 12.367 mm)
Valve spring length	½ in (12.7 mm)
Ball diameter	7/32 in (5.537 mm)
Aluminium crosshead width	0.497 - 0.498 in (12.624 - 12.649 mm)
Working clearance in plunger heads	0.0015 - 0.0045 in (0.038 - 0.11 mm)

Oil pressure release valve

Piston diameter	0.5605 - 0.5610 in (14.236 - 14.239 mm)
Pressure release operates ... ...: ...	60 lb sq in (4.22 kg sq cm)
Spring length (free)	$1^3/8$ in (39.926 mm)
Load at $1^3/16$ in	8 lb (3.629 kg)
Rate	42.3 lb (19.18 kg)

Oil pressure

Normal running	60 lb sq in (4.218 kg cm^2)
Idling	20 - 25 lb sq in (1.406 - 1.687 kg cm^2)

Model T100T — Daytona Sports

Fuel system

Carburettors (two)	Before H 57083 Monobloc	After H 57083 Concentric
Amal type	376/324 and 325	626/9 and 10
Main jet size	200	140
Pilot jet size	25	—
Needle jet size	0.106	0.106
Needle type	C	Two rings above needle clip grooves *
Needle position	3	2
Throttle valve type	376/3½	3
Nominal bore size	$1^1/16$ in (27 mm)	26 mm
Air cleaner type	Coarse felt	Coarse cloth

* An alternative needle is available for use with alcohol, marked Y

Model T100S — Tiger 100

Fuel system

Carburettor:	Before H 57083 Monobloc	After H 57083 Concentric
Amal type	376/273	628/8
Main jet size	190	180
Pilot jet size	25	—
Needle jet size	0.106	0.106
Needle type	C	Two scribed rings above needle clip grooves *
Needle position	3	2
Throttle valve type	376.32	4
Carburettor nominal bore size	1 in	26 mm
Air cleaner type	Felt or paper element	Felt

* An alternative needle is available for use with alcohol, marked Y

Model T90 — Tiger 90

Fuel system

Carburettor	Before H 57083 Monobloc	After H 57083 Concentric
Amal type	376/300	624/2
Main jet size	180	180
Pilot jet size	20	—
Needle jet size	0.106	0.106
Needle type	C	Two scribed rings above needle clip grooves
Needle position	3	2
Throttle type	376/3	3½
Carburettor nominal bore size	$15/16$ in	24 mm
Air cleaner type	Felt or paper element	Felt

Model 5TA — Speed Twin

Carburettor

Amal type	Monobloc 376/273
Main jet size	190
Pilot jet size	25

Needle jet size	0.106
Needle type	C
Needle position	3
Throttle valve type	376/3½
Return spring free length	2½ in (63.5 mm)
Carburettor nominal bore size	1 in (25.4 mm)
Air cleaner type	Felt or paper element

Model 3TA — Twenty One

Carburettor

Type	Monobloc 375/62
Main jet	100
Pilot jet	25
Needle jet	0.106
Needle position	3
Needle type	B
Throttle valve	375/3½
Return spring length	2½ in (63.2 mm)
Nominal bore	25/32 in (19.84 mm)
Air cleaner	Felt or paper element

Unless otherwise stated, R100R data applies to all models

1 General description

The fuel system comprises a petrol tank from which petrol is fed by gravity to the float chamber(s) of the carburettor(s). Two petrol taps, with built-in gauze filter, are located one each side beneath the rear end of the petrol tank. For normal running the right hand tap alone should be opened except under high speed and racing conditions. The left hand tap is used to provide a reserve supply, when the main contents of the petrol tank are exhausted.

For cold starting the carburettor(s) incorporate an air slide which acts as a choke controlled from a lever on the handlebars. As soon as the engine has started, the choke can be opened gradually until the engine will accept full air under normal running conditions.

Lubrication is effected by the 'dry sump' principle in which oil from the separate oil tank is delivered by gravity to the mechanical oil pump located within the timing chest. Oil is distributed under pressure from the oil pump through drillings in the crankshaft to the big ends where the oil escapes and is fed by splash to the cylinder walls, ball journal main bearings and the other internal engine parts. Pressure is controlled by a pressure release valve, also within the timing chest. After lubricating the various engine components, the oil falls back into the crankcase, where it is returned to the oil tank by means of the scavenge pump. A bleed-off from the return feed to the oil tank is arranged to lubricate the rocker arms and valve gear, after which it falls by gravity via the pushrod tubes and the tappet blocks, to the crankcase. An additional, positive oil feed is arranged from drillings in the timing cover to lubricate the exhaust tappets. It will be noted that the oil pump is designed so that the scavenge plunger has a greater capacity than the feed plunger; this is necessary to ensure that the crankcase is not flooded with oil, and that any oil drain-back whilst the machine is standing is cleared quickly, immediately the engine starts.

2 Petrol tank - removal and replacement

1 The petrol tank is secured to the frame by two studs underneath the nose, one on each side. These studs project through two short brackets welded to the frame and are cushioned by rubber washers to damp out vibration. The tank is retained at the front by two self-locking nuts and washers which thread onto the studs. Early models have two bolts, threaded directly into the tank and wire-locked together. The rear mounting takes the form of a lug welded to the rear of the tank which matches with a threaded hole in the top portion of the frame, close to the nose of the dual seat. Anchorage is provided by a bolt which passes through shaped rubbers to provide a flexible mounting.

2 When the bolt and two nuts are removed and the two fuel pipe unions disconnected at their joint with the petrol taps, the tank can be lifted from the machine. Make sure the shaped rubbers are not lost, since they will be displaced as the tank is removed.

3 When replacing the tank, special care must be taken to ensure that none of the carburettor control cables are trapped or bent to a sharp radius. Apart from making control operation much heavier, there is risk that the throttle may stick since there is minimum clearance between the underside of the petrol tank and the top frame tube.

4 Models exported to the USA have reflectors fitted below the front of the petrol tank secured by the front mounting nuts or bolts. The reflector units must be removed first. If necessary the reflector lenses can be prised out of position to gain better access to the nuts or bolts.

3 Petrol taps - removal and replacement

1 The petrol taps are threaded into inserts in the rear of the petrol tank, at the underside. Neither tap contains provision for turning on a reserve quantity of fuel. It is customary to use the right hand tap only so that the left hand tap will supply the reserve quantity of fuel, unless the machine is used for high speed work or racing. In these latter cases, it is essential to use both taps in order to obviate the risk of fuel starvation.

2 Before either tap can be unscrewed and removed, the petrol tank must be drained. When the taps are removed each gauze filter, which is an integral part of the tap body, will be exposed.

3 When the taps are replaced, each should have a new sealing washer to prevent leakage from the threaded insert in the bottom of the tank. Do not overtighten; it should be sufficient just to commence compressing the fibre sealing washer.

4 Petrol feed pipes - examination

1 Plastic feed pipes of the transparent variety are used with a union connection to each petrol tap and a push-on fit at the carburettor float chamber.

2 After lengthy service, the pipes will discolour and harden gradually due to the action of the petrol. There is no necessity to renew the pipes at this stage unless cracks become apparent or the pipe becomes rigid and 'brittle'.

5 Carburettor(s) - removal

1   Both single and twin carburettor fitments have been used, depending on the version. Early models used the Amal Monobloc carburettor(s) whilst later and now current versions use the Amal Concentric carburettor(s). Both types are described here but special emphasis is given to the concentric because it is, by now, the most usual fitment or replacement.
2   Before removing a carburettor it is first necessary to detach the mixing chamber top which is retained by two small screws and lift away the top complete with the control cables, throttle valve and air slide assemblies. The petrol pipe can then be pulled off the push connection at the float chamber (or the union complete detached) and, after detaching the two retaining nuts and shakeproof washers, the complete carburettor may be removed from the cylinder head.

6 Carburettor(s) - dismantling, examination and reassembly

Amal Concentric carburettor only

1   To remove the float chamber, unscrew the two crosshead screws on the underside of the mixing chamber. The float chamber can then be pulled away complete with float assembly and sealing gasket. Remove the gasket and lift out the horseshoe shaped float, float needle and spindle on which the float pivots.
2   When the float chamber has been removed, access is available to the main jet, jet holder and needle jet. The main jet threads into the jet holder and should be removed first, from the underside of the mixing chamber. Next unscrew the jet holder which contains the needle jet. The needle jet cannot be removed until the jet holder has been unscrewed and removed from the mixing chamber because it threads into the jet holder from the top. There is no necessity to remove the throttle stop or air adjusting screws.
3   Check the float needle for wear which will be evident in the form of a ridge worn close to the point. Renew the needle if there is any doubt about its condition, otherwise persistent carburettor flooding may occur.
4   The float itself is unlikely to give trouble unless it is punctured and admits petrol. This type of failure will be self-evident and will necessitate renewal of the float.
5   The pivot needle must be straight - check by rolling the needle on a sheet of plate glass.
6   It is important that the gasket between the float chamber and the mixing chamber is in good condition if a petrol tight joint is to be made. If it proves necessary to make a replacement gasket, it must follow the exact shape of the original. A portion of the gasket helps retain the float pivot in its correct location; if the pin rides free it may become displaced and allow the float to rise, causing continual flooding and difficulty in tracing the cause. Use Amal replacements whenever possible.
7   Remove the union at the base of the float chamber and check that the inner nylon filter is clean. All sealing washers must be in good condition.
8   Make sure that the float chamber is clean before replacing the float and float needle assembly. The float needle must engage correctly with the lip formed on the float pivot; it has a groove which must engage with the lip. Check that the sealing gasket is placed OVER the float pivot spindle and the spindle is positioned correctly in its seating.
9   Check that the main jet and needle jet are clean and unobstructed before replacing them in the mixing chamber body. Never use wire or any pointed instrument to clear a blocked jet, otherwise there is risk of enlarging the orifice and changing the carburation. Compressed air provides the best means, using a tyre pump if necessary.
10 Before refitting the float chamber, check that the jet holder and main jet are tight. Do not invert the float chamber otherwise the inner components will be displaced as the retaining screws are fitted. Each screw should have a spring washer to obviate the risk of slackening.

2.1 Rear tank mounting must have insulating washers fitted correctly

2.4 Front tank mountings carry side reflectors on export models

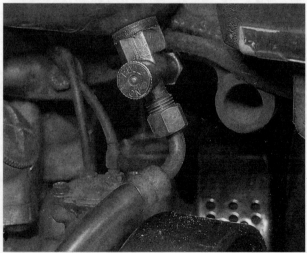

3.1 Petrol taps are of cork push-pull type

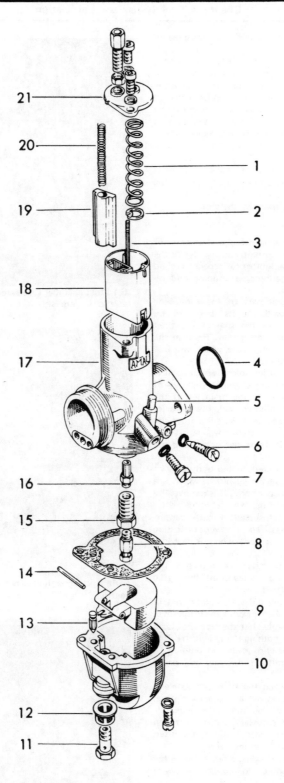

Fig. 4.1. Component parts of the Amal concentric carburettor

1	Throttle return spring	7	Throttle stop screw	13	Float needle	19	Air slide (choke)
2	Needle clip	8	Main jet	14	Float hinge	20	Air slide return spring
3	Needle	9	Float	15	Jet holder	21	Mixing chamber top
4	"O" ring	10	Float chamber	16	Needle jet		
5	Tickler	11	Banjo union bolt	17	Mixing chamber body		
6	Pilot jet screw	12	Filter	18	Throttle valve (slide)		

11 When replacing the carburettor, check that the O ring seal in the flange mounting is in good condition. It provides an airtight seal between the carburettor flange and the cylinder head flange to ensure the mixture strength is constant. Do not overtighten the carburettor retaining nuts for it is only too easy to bow the flange and give rise to air leaks. A bowed flange can be corrected by removing the O ring and rubbing down on a sheet of fine emery cloth wrapped around a sheet of plate glass, using a circular motion. A straight edge will show if the flange is level again, when the O ring can be replaced and the carburettor refitted.

12 Before the mixing chamber top is replaced, check the throttle valve for wear. A worn valve is often responsible for a clicking noise when the throttle is opened and closed. Check that the needle is not bent and that it is held firmly by the clip.

Amal Monobloc carburettor only

13 Early models were fitted with the Amal monobloc carburettor which preceded the concentric type, currently in use. Since the two designs of carburettor differ in a number of respects, revised procedure is necessary when dismantling, examining and re-assembling the monobloc instrument.

14 The float chamber is an integral part of the monobloc carburettor and cannot be separated. Access is gained by removing three countersunk screws in the side of the float chamber and removing the end cover and gasket. Remove the small brass distance piece and the float needle which will free from its seating as the float is withdrawn.

15 The main jet threads into the main jet holder which itself is screwed into the main body of the mixing chamber. Removal of the lower main jet cover gives access to the main jet. If the hexagonal nut above the jet cover is unscrewed, the main jet holder can be detached and the needle jet unscrewed from the upper end. The pilot jet has its own separate cover nut. When removed, the jet can be unscrewed. It is threaded at one end and has a screwdriver slot.

16 Unless internal blockages are suspected, or the body is worn badly, there is no necessity to remove the jet block, which is a tight push fit within the mixing chamber body. It is removed by pressing upward, through the orifice of the main jet holder, after removing the small locating peg which threads into the carburettor body. Extreme care must be exercised to prevent distorting either the jet block or the carburettor body which is cast in a zinc-based alloy.

17 Check the float needle for wear by examining it closely. If a ridge has worn around the needle, close to the point, the needle should be discarded and a new one fitted.

18 The float is unlikely to give trouble unless it is punctured, in which case a replacement is essential. Do not omit to fit the small brass distance piece on the float pivot, after the float has been inserted. If this part is lost, there is nothing to prevent the float moving across to the float chamber end cover and binding - a fault which will give rise to intermittent flooding and prove difficult to pinpoint.

19 There must be a good seal between the float chamber end cover and the float chamber. Always use a new gasket when the seal is broken to obviate the risk of continual petrol leakage.

20 Do not omit to inspect and, if necessary, clean the nylon filter within the float chamber union. When replacing the filter, position it so that the gauze is facing the inflow of petrol. On some of the earlier filters, the plastic dividing strips between the gauze segments are somewhat wide and could impede the flow of petrol under full flow conditions.

21 As stressed in the preceding part of this Section, do not use wire or any pointed object to clear blocked jets. Compressed air should be used to clear blockages; even a tyre pump can be utilised if a compressed air line is not available.

22 The monobloc carburettor has an O ring in the centre of the mounting flange which must be in good condition if air leaks are to be excluded. If the flange is bowed, as the result of previous overtightening, the O ring should be removed and the flange rubbed down on fine emery cloth wrapped around a sheet of plate glass. Rub with a rotary motion and when a straight edge

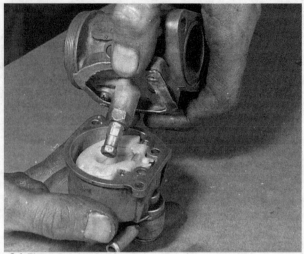

6.1 Float chamber is released by removing two retaining screws on underside

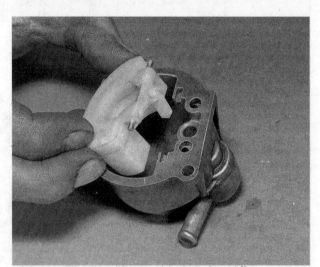

6.1a Horseshoe float lifts out with pivot pin and float needle

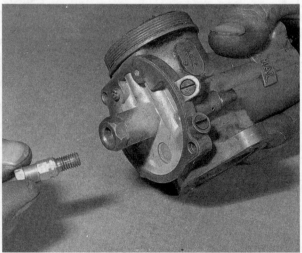

6.2 Main jet threads into jet holder

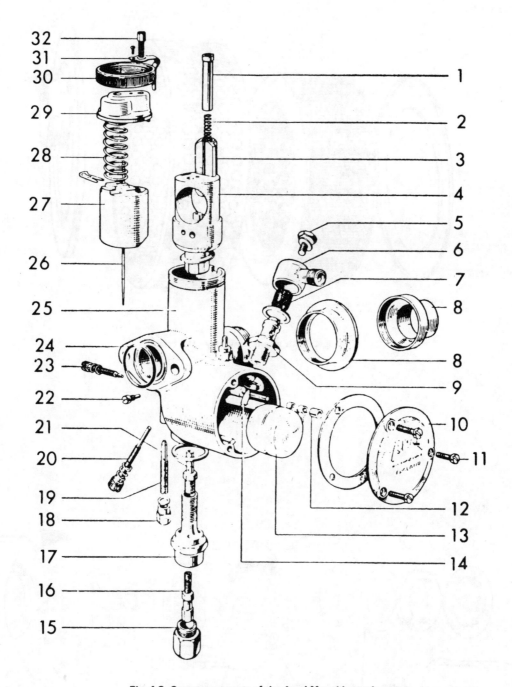

Fig. 4.2. Component parts of the Amal Monobloc carburettor

1 Air valve guide	9 Needle setting	18 Pilot jet cover nut	27 Throttle slide
2 Air valve spring	10 Float chamber cover	19 Pilot jet	28 Throttle spring
3 Air valve	11 Cover screw	20 Throttle stop screw	29 Top
4 Jet block	12 Float spindle bush	21 Needle jet	30 Cap
5 Banjo bolt	13 Float	22 Locating peg	31 Click spring
6 Banjo	14 Float needle	23 Air screw	32 Adjuster
7 Filter gauze	15 Main jet cover	24 "O" ring seal	
8 Air filter connection (top of air intake tube)	16 Main jet	25 Mixing chamber	
	17 Main jet holder	26 Jet needle	

Fig. 4.3. Air cleaner assembly - UK type

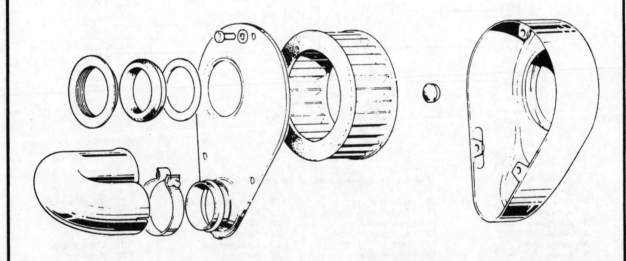

Fig. 4.4. Air cleaner assembly - export type

81

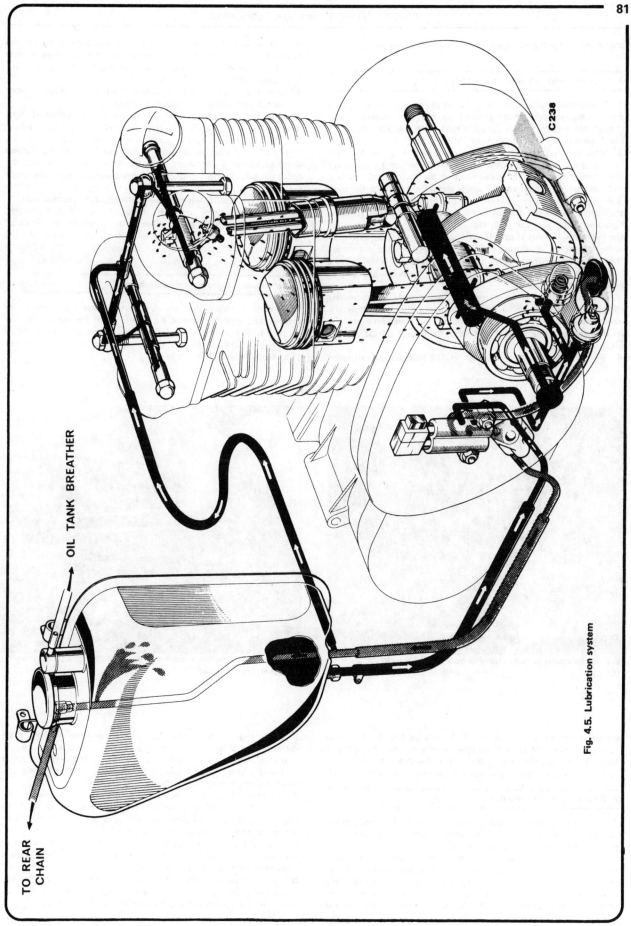

TO REAR CHAIN

OIL TANK BREATHER

C238

Fig. 4.5. Lubrication system

shows the flange is level again, the O ring can be replaced.

7 Carburettor(s) - checking the settings

1   The various sizes of jets and that of the throttle slide, needle and needle jet are predetermined by the manufacturer and should not require modification. Check with the Specifications list if there is any doubt about the values fitted.

2   Slow running is controlled by a combination of the throttle stop and air regulating screw settings. Commence by screwing the throttle stop screw(s) inward so that the engine runs at a fast tickover speed. Adjust the air screw setting(s) until the tickover is even, without either misfiring or 'hunting'. Screw the throttle stop screw(s) outward again until the desired tickover speed is obtained, then recheck with the air regulating screw(s) so that the tickover is as even as possible. Always make these adjustments with the engine at normal running temperature and remember that an engine fitted with high-lift cams is unlikely to run evenly at very low speeds no matter how carefully the adjustments are made.

3   If desired, there is no reason why the throttle stop screw(s) should not be lowered so that the engine will stop completely when the throttle is closed. Some riders prefer this arrangement so that the maximum braking effect of the engine can be utilised on the over-run.

4   As an approximate guide, up to 1/8 throttle is controlled

same time, if they do not, use the cable adjusters to ensure the moment of lift coincides. It is important that the throttle stop screws are slackened off during this operation to obviate the risk of a false reading.

3   Cross-check by noting the points at which the throttle slides lift completely and adjust again, if necessary.

4   Start the engine and when it is at running temperature, stop it and remove one spark plug lead. Restart the engine and adjust the air regulating screw and throttle stop screw of the OPPOSITE cylinder as detailed in Section 7.2 until the desired tickover speed is obtained. Stop the engine again, replace the spark plug lead and repeat the whole operation with the other cylinder and carburettor.

5   When both spark plug leads are replaced, it is probable that the tickover speed will be too high. It can be reduced to the desired level by unscrewing both throttle stop screws an identical amount and rechecking to ensure both throttle valves still lift simultaneously.

9 Air cleaner - removal and replacement

1   Models supplied for the UK market have a circular air cleaner which either screws or pushes onto the carburettor air intake(s). The element is of the corrugated paper type and is removed by slackening the clamping screw of the perforated strip around the air cleaner body.

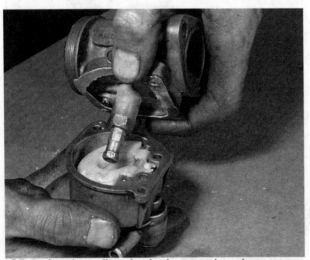

6.7. Banjo union at float chamber base contains nylon filter

8.1 Throttle cable has a one-into-two junction box when twin carburettors are fitted

by the pilot jet, from 1/8 to 1/4 throttle by the throttle valve cutaway, from 1/4 to 3/4 throttle by the needle position and from 3/4 to full throttle by the size of the main jet. These are only approximate divisions; there is a certain amount of overlap.

8 Balancing twin carburettors

1   The twin carburettor cables usually need no attention, as the junction boxes are plastic. No maintenance is necessary other than regular lubrication of the cables.

2   Before starting the balancing operation, it is essential to check that both carburettors operate simultaneously. Place a finger inside the bell mouth of each carburettor intake in turn, and check when the throttle valve commences to move as the twist grip is rotated. Both slides should begin to rise at exactly the

2   Models destined for the USA have a variation of the UK air cleaner, which takes the form of a pear-shaped box attached to the carburettor air intake(s). Access to the element, which is of the dry cloth type, is gained by removing the three hexagon headed bolts through the detachable end plate.

3   To clean either type of element, tap it gently to remove all loose dust, then direct a jet of compressed air on to it. If the element is damp, oil-stained or otherwise blocked, it should be discarded and a new replacement fitted.

4   Do NOT oil either type of filter element or the flow of air to the carburettor will be restricted, resulting in high petrol consumption and poor performance.

5   When replacing the element(s), reverse the dismantling procedure. Do not run the machine without the air cleaners fitted, unless the carburettor(s) is re-jetted to compensate.

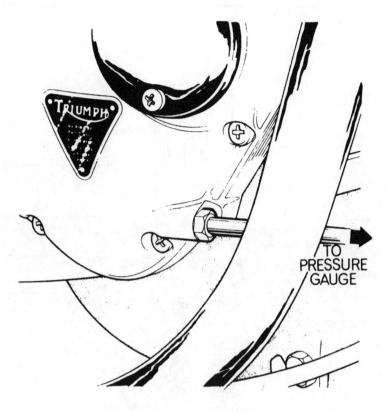

Fig. 4.6. Oil pressure check point

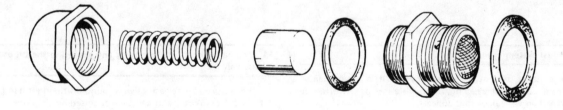

Fig. 4.7. Oil pressure release valve

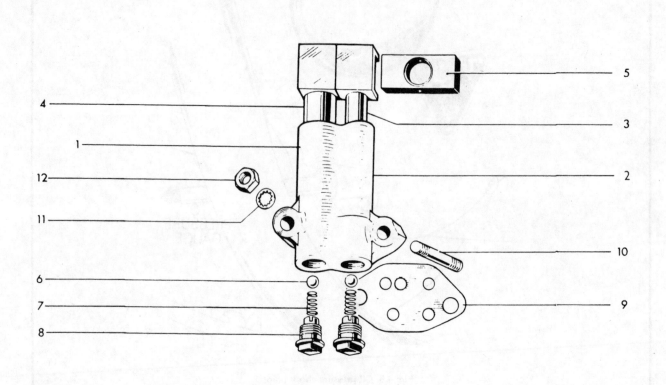

Fig. 4.8. Oil pump

1	Oil pump complete	7	Valve spring - 2 off
2	Pump body	8	Screwed plug - 2 off
3	Feed plunger	9	Oil pump gasket
4	Scavenge plunger	10	Oil pump stud - 2 off
5	Drive block	11	Serrated washer - 2 off
6	Valve ball (7/16 in. - **2** off	12	Nut - 2 off

10 Exhaust system - general

1 The exhaust system(s) usually needs little maintenance, apart from occasional cleaning, although a loss of performance will occur if an air leak or split appears.

2 Air leaks are the most likely to upset the exhaust system and may cause a backfire on the over-run. They usually occur after reassembly and can be attributed to a leak at the exhaust port or silencer joint. Make sure when reassembling the system that all pipes are round and not damaged at the ends where they enter the silencers, all clips must be in good condition, and the bolts free from rust.

3 If all these points are remembered and the system is bolted on firmly, no trouble should occur.

4 Never tamper with the silencers by removing the baffles or the like. They are made to give maximum performance with minimum noise and more often than not (unless tuned properly for racing purposes) more noise means less speed, a poor reward for tampering with them.

11 Engine lubrication - removing and replacing the oil pressure release valve

1 The oil pressure release valve is found on the right hand front half of the crankcase and can be recognised by the large, domed chrome nut which projects it.

2 The pressure release valve rarely gives trouble because of its simplicity. To check the oil pressure, remove the hexagon-headed plug in the forward edge of the timing cover and screw in an oil pipe leading to an oil pressure gauge calibrated up to 100 psi. If the oil pressure is found to be below the normal running rate of 60 psi (4.22 kg sq cm) when the engine is run, and the bearings are known to be good, a check of the valve is advisable.

3 Unscrew the large hexagonal nut abutting the crankcase surface.

4 Unscrew the domed nut to release the spring and piston; check for damage and clean the gauze. **Do not use a vice in this operation as it may cause distortion**. Always clean and oil the

working parts.

5 The one item which can give low oil pressure is the spring, as the tension of the spring dictates the oil pressure; a weak spring means low oil pressure.

6 To check the spring, measure its free length. If uncompressed it should measure 1 3/8 in (34.925 mm).

7 The oil pressure when idling (ticking over speed) should be 20 - 25 psi (1.406 - 1.687 kg sq cm).

8 Reassemble in the reverse order, not forgetting to renew the two fibre washers.

12 Engine lubrication - removal, examination and replacement of the oil pump

1 The oil pump is located within the timing cover, which must be removed to gain access. (Refer to Chapter 1, Section 8.) The oil pump is retained by two conical nuts which should not be removed until the oil tank has been drained. When both nuts are removed, the oil pump is free to be drawn off the mounting studs.

2 The part subject to most wear is the drive block slider, which should be renewed to maintain pump efficiency. Wear will be obvious immediately on examining the block.

3 The oil pump plungers are not so vulnerable as they are permanently immersed in oil. They should, however, be checked for scoring or any undue slackness in the bores. The normal running clearance is up to 0.0005 inch in the case of both the scavenge and feed plungers.

4 Remove the square-headed plugs from the bottom of each plunger housing and check that in each case the ball valve is not sticking to its seat. The springs should have a free length of 1/2 inch; if they have taken a permanent set they must be removed.

5 When reassembling the pump, prime both plunger bores with engine oil and check that oil is forced through the outlet ports as the plungers are inserted and depressed. Check that the oil levels within the bores do not fall as the plungers are raised. If they fall,

this denotes a badly seating ball valve at the base of the plunger. Dismantle the ball valve assembly again and clean the seating. The seating can be restored by tapping the ball on its seating with a punch, **but only if the pump body is made of brass.** Replace the spring and the end cap, then recheck again.

6 Always fit a new gasket when the pump is refitted to the timing chest and check that the oil holes align correctly. Do not use gasket cement of any kind as this may restrict or block the oilways. Check that the drive block slider engages correctly with the peg on the inlet camshaft pinion, before replacing and tightening the conical retaining nuts.

7 Clean off and use new gasket cement on the timing cover when it is replaced, to ensure an oiltight joint. Do not forget to replace the blanking off plug when the oil pressure gauge is removed.

8 Remember that most lubrication troubles are caused by failure to change the engine oil at the prescribed intervals. Oil is cheaper than bearings.

13 Locating and cleaning the engine oil filters

1 The oil in new or reconditioned engines should be changed at 250, 500 and 1000 mile intervals (400, 800, 1500 km) during running in, and thereafter as in the Routine Maintenance schedules.

2 The oil filters are made of gauze and are easily cleaned in paraffin (Kerosene) before they are replaced.

3 When cleaning the oil tank filter, it will be necessary to drain off the oil before taking the filter out of the oil tank. Wash off the gauze and replace the filter in the tank. The oil feed pipe is connected to this filter by a union nut.

4 The other filter is under the crankcase. A small amount of oil will be released when this is taken out. Clean the gauze and refit using a new sealing washer.

5 Make sure that all filter nuts are tight, also the drain plugs. Refill the oil tank with clean oil.

10.2 Exhaust pipe must be a good fit on exhaust stub

10.2a Clamp bolts on balance pipe must be tight

14 Fault Diagnosis

Symptom	Reason/s	Remedy
Excessive fuel consumption	Air filter choked, damp or oily	Check and if necessary renew.
	Fuel leaking from carburettor	Check all unions and gaskets.
	Float needle sticking	Float needle seat needs cleaning.
	Worn carburettor	Renew.
Idling speed too high	Throttle stop screw in too far	Re-adjust screw.
	Carburettor top loose	Tighten top.
Engine does not respond to throttle	Mixture too rich	Check for displaced or punctured float.
Engine dies after running for a short while	Blocked air vent in filler cap	Clean.
	Dirt or water in carburettor	Remove and clean float chamber.
General lack of performance	Weak mixture; float needle stuck in seat	Remove float chamber and check.
	Leak between carburettor and cylinder head	Bowed flange; rub down until flat and replace O ring seal.
	Fuel starvation	Turn on both petrol taps for fast road work.

11.1 Pressure release valve is at front of right-hand crankcase

11.2 Pressure release plunger must be clean and free in housing

12.1 Oil pump is driven off inlet camshaft pinion

12.2 Alloy drive block engages with pump plungers

Chapter 5 Ignition system

Contents

Specifications

Model T100R — Daytona

Ignition timing
Crankshaft position (BTDC):
   Static timing ... ... 14°
   Fully advanced ... ... 38°
Piston position (BTDC):
   Static timing ... ... 0.052 in (1.32 mm)
   Fully advanced ... ... 0.330 in (8.38 mm)
Advance range:
   Contact breaker ... ... 12°
   Crankshaft ... ... 24°

Contact breaker
Gap setting ... ... 0.014 - 0.016 in (35 - 40 mm)
Advance range ... ... 12° (24° crankshaft)
Fully advanced at ... ... 2000 rpm

Spark plugs
Type 				Champion N4
Plug gap settings ... ... 0.020 in (0.5 mm)
Thread size ... ... 14 mm x ¾ in reach

Model T90 — Tiger 90

Ignition timing
Crankshaft position (BTDC):
   Static timing ... ... 14°
   Fully advanced ... ... 38°
Piston position (BTDC):
   Static timing ... ... 0.052 in (1.32 mm)
   Fully advanced ... ... 0.330 in (8.38 mm)
Advance range:
   Contact breaker ... ... 12°
   Crankshaft ... ... 24°

Spark plugs
Type 				Champion N4
Plug gap settings ... ... 0.020 in (0.5 mm)
Thread size ... ... 14 mm x ¾ in reach

Model 5TA — Speed Twin

Ignition timing
   Crankshaft position (BTDC):
      Static timing ... ...   12°
      Fully advanced ... ...   36°
   Piston position (BTDC):
      Static timing ... ...   0.035 in (0.90 mm)
      Fully advanced ... ...   0.310 in (7.93 mm)
   Advance range:
      Contact breaker ... ...   12°
      Crankshaft ... ...   24°

Spark plugs
   Type ... ...   Champion N4
   Plug gap setting ... ...   0.020 in (0.5 mm)
   Thread size ... ...   14 mm

Model 3TA — Twenty One

Ignition timing
   Crankshaft position (BTDC):
      Static timing ... ...   6°
      Fully advanced ... ...   30°
   Piston position (BTDC):
      Static timing ... ...   0.010 in (0.25 mm)
      Fully advanced ... ...   0.210 in (5.33 mm)
   Advance range:
      Contact breaker ... ...   12°
      Crankshaft ... ...   24°

Spark plugs
   Type ... ...   Champion N4
   Plug gap settings ... ...   0.020 in (0.50 mm)
   Thread size ... ...   14 mm x ¾ in reach

Alternative data for Sports models fitted with AC magneto (ET) ignition equipment (prior to H 57083)

Ignition timing
   Crankshaft position (BTDC):
      Static timing ... ...   27°
      Fully advanced ... ...   37°
   Piston position (BTDC):
      Static timing ... ...   0.173 in (4.39 mm)
      Fully advanced ... ...   0.320 in (8.128 mm)
   Advance range:
      Contact breaker ... ...   5°
      Crankshaft ... ...   10°

Contact breaker
   Gap setting ... ...   0.014 - 0.016 in (0.35 - 0.40 mm)
   Advance range ... ...   5°
   Fully advanced at ... ...   2000 rpm

Spark plugs
   Type ... ...: ...   Champion N4
   Plug gap setting ... ...   0.020 in (0.5 mm)
   Thread size ... ...   14 mm x ¾ in reach

     Unless otherwise stated, T100R data applies to all models

1 General description

The spark necessary to ignite the petrol/air mixture in each combustion chamber is derived from a battery and coil, used in conjunction with a contact breaker to determine the precise moment at which the spark will occur. As the points separate the circuit is broken and a high tension voltage is developed across the points of the spark plug which jumps the air gap and ignites the mixture. Each cylinder has its own ignition circuit, hence the need for two separate ignition coils and a twin contact breaker assembly.

When the engine is running, the surplus current produced by the generator is converted into direct current by the rectifier and used to charge the battery. The six volt system used for the earlier models contains provision for emergency starting if the battery is fully discharged. This facility is not required in the case of the twelve volt system because the generator provides sufficient current for the initial start under similar circumstances. Generator output does not correspond directly to engine rpm and is regulated by a diode in circuit, eliminating the need for an electro-mechanical device such as a voltage regulator. All coils in the system are brought into operation only if there is a heavy electrical load when all lights are in use.

2 Checking the generator output

Specialised test equipment of the multi-meter type is essential to check generator output with any accuracy. It is unlikely that the average owner will have access to this type of equipment or instruction in its use. In consequence, if generator performance is suspect, it should be checked by a Triumph agent or an auto-electrical expert.

3 Ignition coils - checking

1 Ignition coil is a sealed unit, designed to give long service without the need of attention. A twin coil system is used on the Triumph unit-construction twins, with the coils mounted one on each side of a bracket that joins the upper and lower frame tubes. It is necessary to remove the petrol tank in order to gain access. This position was later changed to under the seat on oil-in-the-frame models.

2 To check whether a coil is defective, disconnect the spark plug lead from the coil concerned and turn the engine over until the contact breaker points that relate to the coil being tested are closed (check with colour coding of wire). Switch on the ignition and hold the plug lead about 3/16 inch away from the cylinder head. If the coil is in good order, a strong spark should jump the air gap between the end of the plug lead and the cylinder head, each time the points are flicked open.

3 A coil is most likely to fail if the outer casing is compressed or damaged in any way. Fine gauge wire is used for the secondary winding, and this will break easily if subjected to any strain by a damaged casing.

4 Contact breaker adjustment

1 The contact breaker points are located behind the chromium plated cover attached to the forward end of the timing cover by two crosshead screws. Remove both screws and withdraw the

cover.

2 Two types of contact breaker assembly have been used. Machines with engine numbers prior to H 57083 have the Lucas 4CA type containing condensers. Later models utilise the Lucas 6CA assembly, which is more accessible because the condensers have been transferred to another location.

3 In each case, the correct contact breaker gap is 0.015 inch with the points open fully. To adjust the gap on the 4CA unit, slacken the slotted nut that secures the stationary contact point and move the contact either inward or outward until the gap is correct. Tighten the nut and recheck that the setting is still correct. Repeat this procedure for the other set of points.

4 The 6CA unit has the fixed contact point secured by a locking screw which must be slackened first. Adjustment is effected by turning an eccentric screw in the forked end of each contact breaker point assembly, before tightening the locking screw. Recheck that the setting is correct before repeating the procedure for the second set of points. With the 6CA unit, checking whether the points are open fully is simplified by aligning a scribe mark on the end of the contact breaker cam with the nylon heel of the points set, in each case.

5 It is sometimes found that there is a discrepancy between the points gaps of the 6CA unit when the scribe mark of the cam is aligned with the nylon heels. If the discrepancy is greater than 0.003 it is probably caused by cam run-out and can be cured by tapping the cam with a soft metal drift until it seats correctly. Cases have also occurred where the edge of one of the secondary backplates has fouled the cam. Contact between the cam and the backplate can result in the automatic advance unit remaining in the permanently retarded position, so if run-out is evident, either of these two faults should be investigated and remedied.

5 Contact breaker adjustment - removal, renovation and replacement

1 If the contact breaker points are burned, pitted or worn, they must be removed for dressing. If, however, it is necessary to remove a substantial amount of material before the faces can be restored, the points should be renewed.

2 To remove the contact breaker points from the 4CA unit, remove the securing nuts from the condenser terminals. This will free the return springs from the moving contacts, which can be withdrawn from their respective pivot pins. Removal of the slotted nuts will free the fixed contact points.

3 In the case of the 6CA unit, the moving points are removed by unscrewing the nut that secures each low tension lead wire and removing the lead and nylon bush. The return spring and contact point can then be withdrawn from each pivot pin. The fixed points are each retained by two screws which secure them to the backplate assembly.

4 The points should be dressed with an oilstone or fine emery cloth, taking care to keep them absolutely square throughout the dressing operation. If this precaution is not observed, the points will make angular contact with one another when they are replaced, and will burn away rapidly.

5 Replace the points by reversing the dismantling procedure. Take particular care to replace the insulating washers and spacers in their original positions, otherwise the points will be isolated electrically and the ignition system will no longer function.

6 It is important that the points are maintained in a clean condition, especially in machines fitted with provision for emergency start. The emergency start function depends on very small currents passing across the points and if a resistance builds up, the emergency start system will not function.

7 The timing side (right hand) cylinder ignition system is operated by the contact breaker to which the black/yellow lead is attached (left hand set of points, viewed from the open end of the contact breaker housing).

4.1 The Lucas 6CA contact breaker

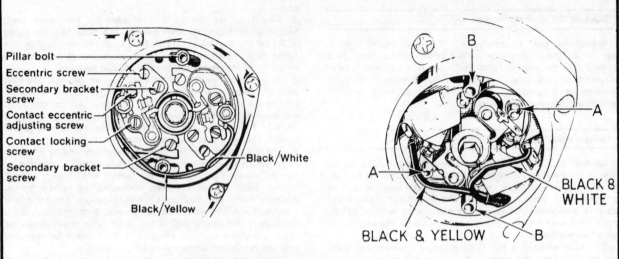

Pillar bolt
Eccentric screw
Secondary bracket screw
Contact eccentric adjusting screw
Contact locking screw
Secondary bracket screw

Black/White

Black/Yellow

B

A

A

BLACK 8 WHITE

BLACK & YELLOW

B

Fig. 5.1. Lucas contact breakers

Contact breaker 4CA
To adjust contact breaker gaps slacken sleeve nuts 'A'.
To rotate contact breaker base plate for setting ignition
timing slacken pillar bolts 'B'.

Fig. 5.2. Setting the contact breaker gaps

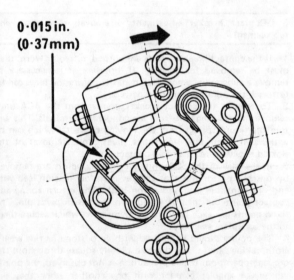

0·015 in.
(0·37mm)

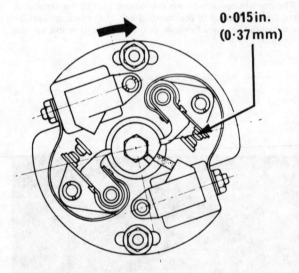

0·015 in.
(0·37mm)

Setting the contact breaker point gap for the right cylinder (black/yellow) lead, illustrating position of the cam where points are just fully open.

Setting the contact breaker point gap for the left cylinder (black/white) lead, illustrating the second position of the cam, where the points have just achieved the fully open position

6 Condensers - removal and replacement

1 As mentioned in the preceding section, the condensers used in conjunction with the Lucas 4CA contact breakers assembly are contained within the contact breakers housing, attached to the backplate. It is advisable to withdraw the assembly complete before detaching the condensers, by removing the two pillar bolts securing the backplate to the timing cover housing.

2 When the Lucas 6CA contact breaker assembly is fitted, the condensers are located remotely. They are attached to a plate suspended from the bracket to which the front petrol tank mountings are bolted, immediately below the nose of the petrol tank.

3 If the engine becomes difficult to start, or if misfiring occurs, it is probable that a condenser has failed. Note that it is rare for both condensers to fail simultaneously, unless they have been damaged in an accident. Examine the contact breaker points whilst the engine is running to see whether arcing is taking place and, when the engine is stopped, examine the faces of the points. Arcing taking place or the points having a blackened and burnt appearance is characteristic of condenser failure.

4 It is not possible to check the condenser without the necessary test equipment. It is therefore best to fit a replacement condenser and observe the effect on engine performance, especially in view of the low cost of the replacement.

7.4 Use of the crankcase timing plug to find TDC

7 Ignition timing - checking and resetting

1 One of two methods can be used to check and if necessary reset the ignition timing, depending on whether a stroboscope is available or whether the static method has to be substituted in its absence. Although the stroboscope method is undoubtedly the more accurate, it is unlikely that many owners will have access to a stroboscope and the various attachments. In consequence, both methods are described in detail, so that the user of this workshop manual will have the option of following whichever the more appropriate.

2 If the static timing method is the one selected, commence by removing all four rocker box inspection caps (or finned inspection covers) and both spark plugs. Place the machine on the centre stand and engage top gear, so that the engine can be rotated by means of the rear wheel.

3 Arrange the right hand cylinder so that the piston is at top dead centre (TDC) on the compression stroke. It is best to use either a dial gauge pressing on the piston crown or a degree disc and adaptor shaft attached to the centre of the exhaust camshaft (Triumph service tool D605/8 and nut S1-51) to maintain absolute accuracy of setting. In this latter case, a timing stick must be used to zero the timing disc accurately. Insert the stick through the right hand spark plug hole and make a mark on both the stick and an adjoining portion of the engine that coincide exactly when the piston is at TDC. Attach a pointer to some convenient part of the engine and rotate the timing disc, INDEPENDENT OF THE ENGINE, so that the zero mark aligns with the pointer. Make a second mark on the timing stick about 1 inch above the original, re-insert the stick and rotate the engine with timing disc attached until this second mark corresponds exactly with the mark originally made on the engine. Take a reading from the timing disc, at the pointer. Now rotate the engine in the opposite direction until the second mark on the timing stick again aligns with the mark on the engine, and take a further reading from the timing disc, at the pointer. If the two readings agree, the timing disc is zeroed correctly and the piston is exactly at TDC. Note that the timing stick should have a rubber plug or some similar device attached to prevent it from falling into the engine if the engine is inadvertently rotated too far in one direction.

4 This procedure can be short cut. These later engines have a plug on top of the crankcase, immediately to the rear of the cylinder barrel, which when removed, exposes the rim of the centre flywheel. If Triumph workshop tool D571/2 is screwed into the hole vacated by the plug, the inner plunger will drop into a hole drilled in the flywheel rim, when the piston is exactly at top dead centre. If the workshop tool is not available, the same effect can be achieved by placing a small socket spanner in the hole and using the shank of a drill as the plunger.

5 Check with the specifications section of this Chapter to ascertain the correct fully-advanced ignition setting recommended for the machine. Note the range of the automatic advance mechanism that is stamped on the REAR of the assembly, double this figure and subtract it from the full advance setting recommended for the machine. This is the correct STATIC setting for the engine. Convert this figure (expressed in degrees) to the equivalent piston displacement before TDC, if a timing stick is used WITHOUT a degree disc (see accompanying table). If a degree disc is used, the static setting in degrees can be used without need for conversion.

6 Rotate the engine BACKWARDS to beyond the setting, then turn it forward again until the correct setting is achieved. This procedure is necessary to take up any backlash that may otherwise affect the accuracy of the setting. Since the left hand set of contact breaker points (black and yellow lead) corresponds with the right hand cylinder, maintain the engine in the predetermined position and rotate the backplate of the contact breaker assembly (after slackening the pillar bolts) until the points are just on the point of separation. If the ignition is switched on during this operation, the exact point can be determined when the ammeter reading kicks back to zero. Tighten the pillar bolts.

7 Rotate the engine through 360 degrees, so that the left hand piston is now at TDC on the compression stroke. Check whether the right hand contact breaker points (black and white lead) are now in a similar position ie about to separate. If not, the accuracy of the second set of points must be corrected by adjusting the points gap. Minor adjustments are permissible; if more than 0.003 in adjustment is necessary, check for run-out as described in Section 4.5.

8 If a stroboscope is available, fit the timing shaft adaptor and timing disc to the exhaust camshaft and set the right hand piston to top dead centre as described in paragraphs 3 and 4 of this Section. Connect the stroboscope to the right hand plug lead, after the spark plugs and rocker inspection caps have been replaced, and start the engine. Note the reading from the timing disc by shining the stroboscope light on the pointer, revving the engine until full auto-advance is obtained. If the reading is not

correct, adjust the contact breaker backplate on its slots (after slackening the pillar bolts) until the correct reading is achieved. Repeat this procedure for the left hand cylinder and make any adjustments by varying the contact breaker gap and NOT the backplate.

9   Before disconnecting the stroboscope and timing disc, a check can be made at lower engine speeds to verify whether the full range of ignition advance is achieved with both cylinders. It should be emphasised that the stroboscope method of engine timing ensures both plugs fire at exactly similar piston movements, thereby ensuring optimum engine running conditions and minimum vibration.

10 Machines having early engine numbers require a slightly different technique for stroboscope timing, although the setting up procedure is identical with that given in paragraph 4. These machines have an inspection plate fitted to the primary chaincase cover, retained by three screws. When the screws are withdrawn and the cover lifted away, it will be observed that there is a mark on the face of the generator rotor which should coincide with a pointer on the chaincase if the ignition timing setting is correct. If the machine does not have a chaincase pointer, it is necessary to fit temporarily a special timing plate (Triumph part number D2014) in its place. The plate has two markings; only the line marked 'B' is applicable to the larger capacity models.

11 Connect the stroboscope to the right hand spark plug lead and start the engine. Shine the stroboscope lamp on the rotor marking in the vicinity of the chaincase pointer or plate with the engine running at more than 2000 rpm. Adjust the contact breaker backplate until the marks coincide, when the timing of the right hand cylinder is correct. Connect the stroboscope to the left hand cylinder and repeat the procedure, in this case turning the eccentric screw in the right hand set of contact breaker points until the marks coincide.

12 When the timing has been verified as correct, remove the various attachments and replace the end cover of the contact breaker assembly, using a new gasket. If the stroboscope method of timing has been used, remove the timing plate from the primary chaincase and replace the circular cover. Replace the rocker box inspection caps or the finned inspection covers.

13 Note that if a 6 or 12 volt stroboscope is used, the power must be from an external source. If the machine's battery is used, there is risk of AC pulses developing in the low tension circuit that may cause the stroboscope to give a false reading.

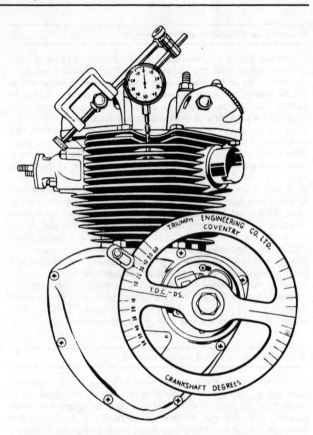

Fig. 5.3. Use of dial gauge and timing disc for static ignition timing

8 Removing and replacing the distributor

1   Early models use a distributor for the ignition system in place of the contact breaker unit within the timing cover. Adjustment of the timing is carried out by slackening the distributor clamp and turning the distributor to either advance or retard the timing.

2   Slacken the clamp bolt at the base of the distributor and pull the complete unit from its housing after disconnecting the lead wires.

3   Check for wear in the distributor shaft bush. If this bush is worn, it must be renewed, otherwise it will be impossible to achieve the correct timing. Service work of this nature is best entrusted to a Triumph specialist or someone experienced in auto-electrical repairs.

4   If the engine has been rotated after the distributor has been removed, it must be aligned correctly before the distributor is replaced. Turn the crankshaft so that the right hand piston is at top dead centre, on the compression stroke. With the clamp plate free enough to permit movement of the distributor, hold the distributor so that when it is viewed downwards the hole for the clamp anchor plate is in the nine o'clock position. Set the distributor so that the contact breaker points are just separating and insert it in its housing. Tighten the clamp bolt. The engine is now ready for accurate timing, which should be accomplished with a degree disc as described in the preceding Section.

Crankshaft position (B.T.D.C.)	Piston position (B.T.D.C.)	
Degrees	in.	mm.
7	·010	·25
8	·015	·38
9	·020	·51
10	·025	·64
11	·030	·76
12	·035	·89
13	·040	1·02
14	·048	1·22
15	·055	1·40
16	·060	1·52
17	·070	1·78
18	·080	2·03
19	·090	2·29
20	·100	2·54
21	·110	2·79

Fig. 5.4. Conversion chart: Engine degree to relative piston positions

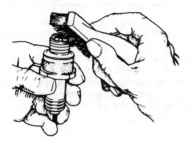

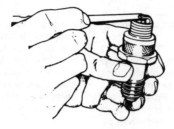

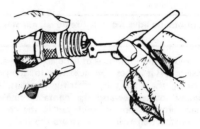

Cleaning deposits from electrodes and surrounding area using a fine wire brush

Checking plug gap with feeler gauges

Altering the plug gap. Note use of correct tool

Fig. 5.5a Sparking plug maintenance

White deposits and damaged porcelain insulation indicating overheating

Broken porcelain insulation due to bent central electrode

Electrodes burnt away due to wrong heat value or chronic pre-ignition (pinking)

Excessive black deposits caused by over-rich mixture or wrong heat value

Mild white deposits and electrode burnt indicating too weak a fuel mixture

Plug in sound condition with lig   greyish brown deposits

Fig. 5.5b. Sparking plug electrode conditions

9 Automatic advance unit - removal, examination and replacement

1   Fixed ignition timing is of little advantage as the engine speed increases and it is therefore necessary to incorporate a method of advancing the timing by centrifugal means. A balance weight assembly located behind the contact breaker, linked to the contact breaker cam, is employed in the case of the Triumph unit-construction twins. It is secured to the exhaust camshaft by a bolt that passes through the centre of the contact breaker cam. It can be withdrawn without need to remove the timing cover, if the contact breaker assembly is removed first.

2   When the assembly is removed from the machine, it is advisable to make a note of the degree figure stamped on the back of the cam unit. This relates to the ignition advance range and, as the previous section has indicated, must be known for accurate static timing.

3   The unit is most likely to malfunction as the result of condensation, which will cause rusting to take place. This will immediately be evident when the assembly is removed. Check that the balance weights move quite freely and that the return springs are in good order. Before replacing the assembly by reversing the dismantling procedure, lubricate the balance weight pivot pins and the cam spindle, and place a light smear of grease on the face of the contact breaker cam. Lubricate the felt pad that bears on the contact breaker cam.

10 Ignition cut-out

1   The ignition circuit is controlled by the ignition switch on the left hand top fork cover. The key can only be withdrawn from the switch when it is in the 'OFF' position and the ignition circuit broken.

2   A state of emergency can occur when the machine is on the move and it is not convenient to reach out for the switch key. Late models have an additional cut-out button on the left hand side of the handlebars, which will break the ignition circuit all the time it is depressed. On these models, the separate lighting switch is mounted in the headlamp shell. Because the ignition circuit is broken only when the cut-out button is depressed, it is essential to turn the switch to the off position and remove the key when the machine is parked.

11 Spark plugs - checking and resetting the gap

1   Each cylinder has a 14 mm spark plug fitted. Each plug must be of the correct heat range. The heat range is denoted by the reference number on the plug. Only the right plug with the correct heat range is good enough for your engine.

2   Do not accept a different plug unless it is the direct equivalent in another manufacturer's range. Each manufacturer has his own number for the same heat range.

3   **Remember that fitting the wrong grade of plug can seriously damage your engine.**

4   The plug should be cleaned and gapped every 2000 miles. To reset the gap bend the outer electrode until a 0.025 in feeler gauge can just be inserted.

5   Never bend or damage the inner electrode, or engine damage may occur.

6   Never overtighten a spark plug. Although a stripped thread can be reclaimed by a Helicoil insert, this means a complete top-end strip.

7   Always carry a spare set of plugs, kept in dry conditions.

8   Make sure that your suppressor caps are not breaking down and affecting television or radio reception. If they are, renewal is necessary.

12 Fault diagnosis

Symptom	Reason/s	Remedy
Engine will not start	No spark at plug	Check whether contact breaker points are opening and also whether they are clean. Check wiring for break or short circuit.
Engine fires on one cylinder	No spark at plug or defective cylinder	Check as above, then test ignition coil. If no spark, see whether points arc when separated. If so, renew condenser.
Engine starts but lacks power	Automatic advance unit stuck or damaged	Check unit for freedom of action and broken springs.
	Ignition timing retarded	Verify accuracy of timing. Check whether points gaps have closed.
Engine starts but runs erratically	Ignition timing too far advanced	Verify accuracy of timing. Points gaps too great.
	Spark plugs too hard	Fit lower grade of plugs and re-test.

Chapter 6 Frame and Forks

Contents

Specifications

Model T100R — Daytona

Telescopic fork

Type	Telescopic with oil damping shuttle valve
Spring details:	
Free length	9¾ in (247.65 mm)
No of working coils	12½
Spring rate	26½ lb in (1.79 kg cm^2)
Colour code	Yellow/blue

	Top bush	Bottom bush
Bush details:		
Length	1 in (25.4 mm)	0.870 - 0.875 in (22.098 - 22.23 mm)
Outer diameter	1.498 - 1.499 in (38.04 - 38.06 mm)	1.4935 - 1.4945 in (37.94 - 37.96 mm)
Inner diameter	1.3065 - 1.3075 in (33.19 - 33.21 mm)	1.2485 - 1.2495 in (31.71 - 31.74 mm)
Stanchion diameter	1.3025 - 1.3030 in (33.08 - 33.096 mm)	
Working clearance in top bush	0.0035 - 0.0050 in (0.0889 - 0.127 mm)	
Fork leg bore diameter	1.498 - 1.500 in (38.04 - 38.1 mm)	
Working clearance of bottom bush	0.0035 - 0.0065 in (0.0889 - 0.165 mm)	
Shuttle valve:		
Outer diameter (large)	1.018 - 1.106 in (24.046 - 28.09 mm)	
Outer diameter (small)	0.875 - 0.874 in (22.27 - 22.205 mm)	

Model T100C — Trophy 500

Front forks

Main spring - load	32 lb in (0.4 kg m)
Colour identification	Yellow/green

Head races

No of balls:	
Top	24
Bottom	24
Ball diameter	3/16 in (4.763 mm)

Swinging fork

Bush type	Phosphor bronze strip
Bush bore diameter	0.8745 - 0.8750 in (21.91 - 21.93 mm)
Spindle diameter	0.8735 - 0.8740 in (22.19 - 22.2 mm)
Distance between fork ends	$7^7/16$ in (189 mm)

Rear suspension

Type	Swinging fork controlled by combined spring/hydraulic damper units. (Bolted up after H 49833)

Spring details:

Fitted length	8 in (203.2 mm)
Free length	$8^3/16$ in (207.96 mm)
Mean coil diameter	1¾ dia (44.45 mm)
Spring rate	145 lb in (10.19 kg cm)
Colour code	Blue/yellow
Load at fitted length	38 lb (2.67 kg)

Model 5TA — Speed Twin

Spring details:

Free length	9¾ in (247.65 mm)
No of working coils	12½
Spring rate	110 lb in (7.73 kg cm)
Colour code	Red/Red

Rear suspension

Spring details:

Fitted length	8 in (203.2 mm)
Free length	$8^3/16$ in (207.96 mm)
Mean coil diameter	1¾ in (44.45 mm)
Spring rate	145 lb in (10.19 kg cm)
Colour code	Blue/yellow
Load at fitted length	38 lb (17.24 kg)

Unless otherwise stated, T100R data applies to all models

1 General description

A full cradle frame is fitted to the Triumph unit-construction twins, in which the front down tube branches into two duplex tubes at the lower end which form the cradle for the unit-construction engine/gear unit.

Rear suspension is provided by a swinging arm assembly that pivots from a lug welded to the vertical tube immediately to the rear of the gearbox. Movement is controlled by two hydraulically-damped rear suspension units, one on each side of the sub-frame. The units have three-rate adjustment, so that the spring loading can be varied to match the conditions under which the machine is to be used.

Front suspension is provided by telescopic forks of conventional design.

2 Front forks - removal from frame

1 It is unlikely that the front forks will need to be removed from the frame as a complete unit unless the steering head bearings require attention or the forks are damaged in an accident.

2 Commence operations by placing the machine on the centre stand and disconnecting the front brake. If the split pin and clevis pin through the U shaped connection to the brake operating arm are removed, the cable can be pulled clear of the cable stop and removed from the cable guide attached to the front mudguard.

3 To remove the front wheel, unscrew and remove the two bolts securing the lower half of each split clamp to the bottom of the fork legs (two nuts, late models). When these clamps are withdrawn, the wheel will drop clear, complete with brake plate and wheel spindle. It may be necessary to raise the front end of the machine a little, in order to gain sufficient clearance for the wheel to clear the mudguard when it is removed.

4 It is convenient at this stage to drain the fork legs of their oil content, if the fork legs are to be dismantled at a later stage. The drain plug is found above the wheel spindle recess on each fork leg. Remove both drain plugs and leave the forks to drain into some suitable receptacle whilst the dismantling continues.

5 There is no necessity to remove the front mudguard unless the fork legs are to be dismantled. The lower mudguard stays bolt direct to lugs at the lower end of each fork leg; the centre fixing is made to the inside of each fork leg where a shaped lug accepts the cut-out of the mudguard stay assembly, which is then retained by a bolt and washer. If the various nuts, bolts and washers are removed, the mudguard can be withdrawn, complete with stays.

6 Detach the headlamp after disconnecting the battery leads. Commence by slackening the screw at the top of the headlamp rim, which will allow the rim and reflector unit to be removed. Detach the pilot bulb holder and pull the snap connectors from the main bulb holder. Disconnect the four spade terminals from the lighting switch and from the ammeter. On earlier models, snap connectors at the wiring harness form an alternative means of disconnection. Disconnect the snap connectors in the warning light leads, then withdraw the wiring harness complete from the headlamp shell. Note that the harness will bring with it the warning light bulb holders and rubber grommets.

7 Remove the pivot bolts on each side of the headlamp shell and withdraw the shell, complete with spacers. A slightly different procedure is necessary in the case of models fitted with flashing indicators, the arms of which pass through the headlamp shell to form the pivots. In this case it is necessary to disconnect the indicator leads and unscrew the nuts around the extensions of the indicator arms which project through the headlamp shell to hold it in position.

8 Detach the control cables from the handlebar controls, or the controls themselves, complete with cables. Remove the dip-switch, the cut-out button (if fitted) and the horn push. Remove the ignition switch, retained in the left hand top fork cover extension by a locknut. The handlebars can now be removed by unscrewing the two eye bolts underneath the top yoke of the

forks, or by removing the split clamps if the handlebars are not rubber-mounted.

9 Remove the steering damper knob and rod (if fitted) by unscrewing the knob until the rod is released from the lower end into which it threads. Slacken the pinch bolt through the top fork yoke, found at the rear of the steering assembly, above the tank. Unscrew the fork stem sleeve nut (if steering damper fitted) or the blind nut (domed) at the top of the steering head column. Slacken and remove the two plated nuts at the top of each fork leg that pass through the top fork yoke.

10 Remove the top fork yoke by tapping on the underside with a rawhide mallet. The forks should be supported throughout this operation because they will free immediately the yoke clears the tapers of the fork inner tubes or stanchions. When the yoke is displaced and removed, the complete fork assembly can be lowered from the steering head and drawn clear. Note that as the head races separate the uncaged ball bearings will be released. Arrangements should be made to catch the ball bearings as they drop free; most probably only those from the lower race will be displaced.

11 It is possible to remove the fork legs separately, if there is no reason to disturb either the steering head assembly or the fork yokes. In this case the plated top fork nuts should be removed and the pinch bolts through each side of the lower fork yoke slackened and removed. Triumph service tool Z19 for machines with engine numbers prior to H57083 should then be threaded into the top of each stanchion to the full depth of thread and used as a drift to drive each stanchion taper free, to release the stanchion complete with lower leg, so that it can be passed through the lower fork yoke and withdrawn from the machine. It is often necessary to open up the pinch bolt joint in the lower fork yoke to prevent the stanchion from binding. If the stanchion has rusted, this will impede its progress through the bottom yoke. Remove all surface rust with emery cloth, wipe clean and apply a light coating of grease or oil.

12 It must be emphasised that the use of the recommended Triumph service tool is essential for this operation. The stanchions are a very tight fit and any attempt to free them with a punch or drift will invariably damage or distort the internal threads, necessitating replacement. An old top fork nut can sometimes be used successfully as a substitute but only if the stanchion is not a particularly tight fit, in either of the fork yokes.

3 Front forks - dismantling and examination

1 When the legs have been removed from the machine, withdraw the fork top covers, or, in the case of some of the early models, the nacelle bottom covers. Remove the cork seating washers. If rubber gaiters are fitted, the top and bottom retaining clips should be slackened and the gaiters removed, together with the clips. Remove the spring abutments and the fork springs.

2 Removal of the dust excluder sleeves which contain a plain washer and oil seal is facilitated by the use of Triumph service tool D220, which engages with the peg holes in the outer surface. If the service tool is not available, a strap spanner or careful work with a centre punch will provide an alternative solution. The fork leg should be supported in a vice during this operation, by clamping the wheel spindle recess. A sharp knock is needed to free the sleeve initially; thereafter it can be unscrewed and removed.

3 Withdraw the stanchion from each leg by withdrawing it whilst the fork leg is still clamped in the vice. A few sharp pulls may be necessary to release the top bush which is a tight fit in the fork leg.

4 Prior to engine number H57083, a different type of internal hydraulic damper assembly was fitted. Before the stanchions can be freed from the fork legs, it is necessary to first unscrew the hexagon headed bolt counter bored into the wheel spindle recesses.

5 Damper units fitted to forks employ an oil restrictor rod assembly, secured by a bolt counter bored into the front wheel spindle recesses of the lower fork legs. Unlike the design that

2.3 Remove split clamps to free wheel spindle

2.3a Don't forget to disconnect the brake cable before wheel is removed

2.5 Lower mudguard stay bolts to ends of fork legs

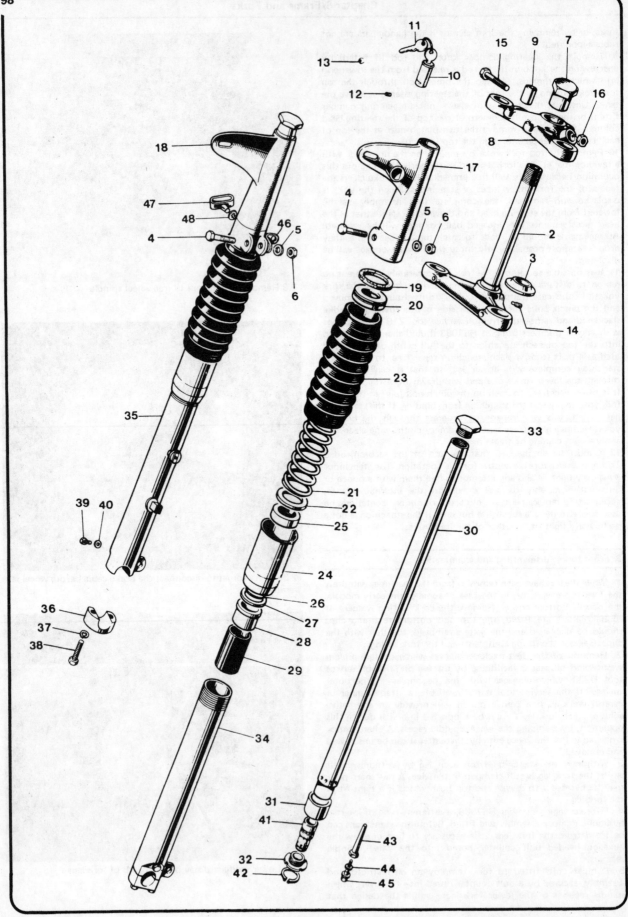

Fig. 6.1. Telescopic front forks

1	Fork assembly 1 off	25	Oil seal 2 off
2	Middle lug and stem 1 off	26	"O" ring 2 off
3	Bottom cone 1 off	27	Plain washer 2 off
4	Pinch bolt 2 off	28	Top bearing 2 off
5	Plain washer 2 off	29	Damping sleeve 2 off
6	Cleveloc nut 2 off	30	Stanchion 2 off
7	Fork stem sleeve nut 1 off	31	Lower bearing 2 off
8	Top lug 2 off	32	Bearing nut 2 off
9	Bonded bush 2 off	33	Cap nut 2 off
10	Lock c/w 2 keys 1 off	34	Left bottom member c/w cap 1 off
11	Key (state serial number) 1 off	35	Right bottom member c/w cap 1 off
12	Grub screw 1 off	36	Spindle cap 2 off
13	Sealing washer 1 off	37	Spring washer 4 off
14	Locating pin 1 off	38	Cap bolt 4 off
15	Pinch bolt 1 off	39	Drain plug 2 off
16	Seated nut 1 off	40	Sealing washer 2 off
17	Left top cover 1 off	41	Shuttle valve 2 off
18	Right top cover 1 off	42	Circlip 2 off
19	Cork washer 2 off	43	Restrictor 2 off
20	Spring abutment 2 off	44	Aluminium washer 2 off
21	Main spring (yellow/blue) 2 off	45	Flanged bolt 2 off
22	Plain washer 2 off	46	Bracket 1 off
23	Telescopic gaiter 2 off	47	Cable retainer 1 off
24	Dust excluder sleeve nut 2 off	48	Starlock washer 1 off

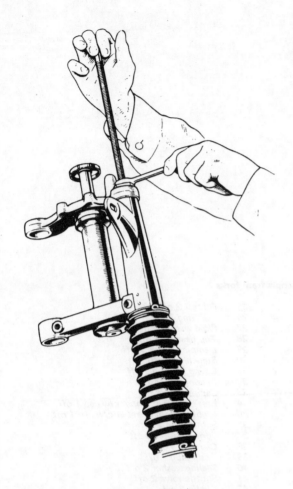

Fig. 6.2. Use of Triumph Service Tool for raising fork stanchions

2.11 Remove pinch bolt to withdraw fork leg

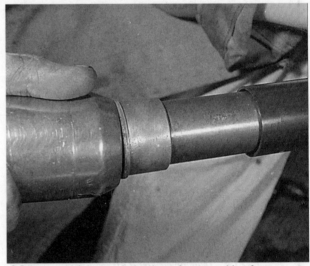

3.2 Removal of dust excluder sleeve frees stanchion from lower fork leg

2.9 Bolts in top of fork legs act also as filler caps

3.3 Give a sharp pull to free stanchion

followed, it is not necessary to detach this assembly before the stanchions can be withdrawn from the lower legs.

6   A further change of damper unit occurred with forks fitted to engine number H57083 onward. These forks are fitted with a shuttle valve damper, attached to the lower end of the fork stanchions. Each shuttle valve is retained by a sleeve nut which also, holds the button bearing of the fork leg. A circlip in front of this nut prevents the valve from passing into the stanchion. This type of damper assembly is easily recognisable by the eight oil bleed holes above the bottom bearing location.

7   The parts most liable to become damaged in an accident are the fork stanchions, which will bend on heavy impact. To check for misalignment, roll the stanchion on a sheet of plate glass, when any irregularity will be obvious immediately. It is possible to straighten a stanchion that has bowed not more than 5/32 in out of true but it is debatable whether this action is desirable. Accident damage often overstresses a component and because it is not possible to determine whether the part being examined has suffered in this way, it would seem prudent to renew rather than repair.

8   Check the top and bottom fork yokes which may also twist or distort in the event of an accident. The top yoke can be checked by temporarily replacing the stanchions and checking whether they lay parallel to one another. Check the lower fork yoke in the same manner, this time with the stanchions inserted until about 6½ inches protrude. Tighten the pinch bolts before checking whether the stanchions are parallel with one another. The lower yoke is made of a malleable material and can be straightened without difficulty or undue risk of fracture.

9   It is possible for the lower fork legs to twist and this can be checked by inserting a dummy wheel spindle made from 11/16 inch diameter bar and replacing the split retaining clamps. If a set square is used to check whether the fork leg is perpendicular to the wheel spindle, any error is readily detected. Renewal of the lower fork leg is necessary if the check shows misalignment.

10   The fork bushes can be checked by positioning the top bush close to the lower bush at the bottom of the stanchion and inserting the assembly in the lower fork leg. Any undue play will immediately be evident, necessitating renewal of the bushes. If the forks are fitted with the older type of grey sintered iron bushes they should be replaced by the later sintered bronze bushes, to gain the benefit of a reduced rate of wear.

11   Examine both fork springs and check that they are of the same length and have not compressed. The springs fitted to all models up to 1962 are 18.25 in (463.55 mm) long when new and those fitted to 1963 models are 17.06 in (433.32 mm) long; both springs must be renewed if either is ½ in (12.7 mm) or more, shorter than the new length. All later models use external fork springs which are 8.75 in (222.25 mm) long on 1964 models and 9.75 in (247.65 mm) long on all other models (1965 on); both springs must be renewed if either is ¼ in (6.35 mm) or more, shorter than the new length. Ensure that both springs are of the same colour-coding.

4   Front forks - examination of the steering head races

1   If the steering head races have been dismantled, it is advisable to examine them prior to reassembling the forks. Wear is usually evident in the form of indentations in the hardened cups and cones, around the ball track. Check that the cups are a tight fit in the steering column headlug.

2   If it is necessary to renew the cups and cones, use a drift to displace the cups by locating with their inner edge. Before inserting the replacements, clean the bore of the headlug. The replacement cups should be drifted into position with a soft metal drift or even a wooden block. To prevent misalignment, make sure that the cups enter the headlug bore squarely. The lower cone can be levered off the bottom fork yoke with tyre levers; the upper cone is within the top fork yoke and can be drifted out. Clean up any burrs before the new replacements are fitted. A length of tubing which will fit over the head stem can be

3.5 Circlip prevents shuttle valve from passing into stanchion

3.9 Lower bush is retained by sleeve nut

used to drive the lower cone into position so that it seats squarely.

3   When the cups and cones are replaced, discard the original ball bearings and fit a new set. It is false economy to re-use the originals in view of the very low renewal cost. Forty ¼ inch diameter balls are required, 20 for each race. Note that when the bearing is assembled, the race is not completely full. There should always be space for one bearing, to prevent the bearings from skidding on one another and wearing more rapidly. Use thick grease to retain the ball bearings in place whilst the forks are being offered up.

5   Front forks - reassembling the fork legs

1   The fork legs are reassembled by following the dismantling procedure in reverse. Make sure all of the moving parts are lubricated before they are assembled. Fit new oil seals, regardless of the condition of the originals.

2   Before refitting the fork stanchions, make sure the external surfaces are clean and free from rust. This will make fitting into the fork yokes at a later stage much easier. Oil, or lightly grease, the outer surfaces after removing all traces of the emery cloth or other cleaner used.

3   When refitting the fork gaiters (if originally fitted) check that they are positioned correctly. The small hole near the area where the clip fastener is located should be at the bottom.

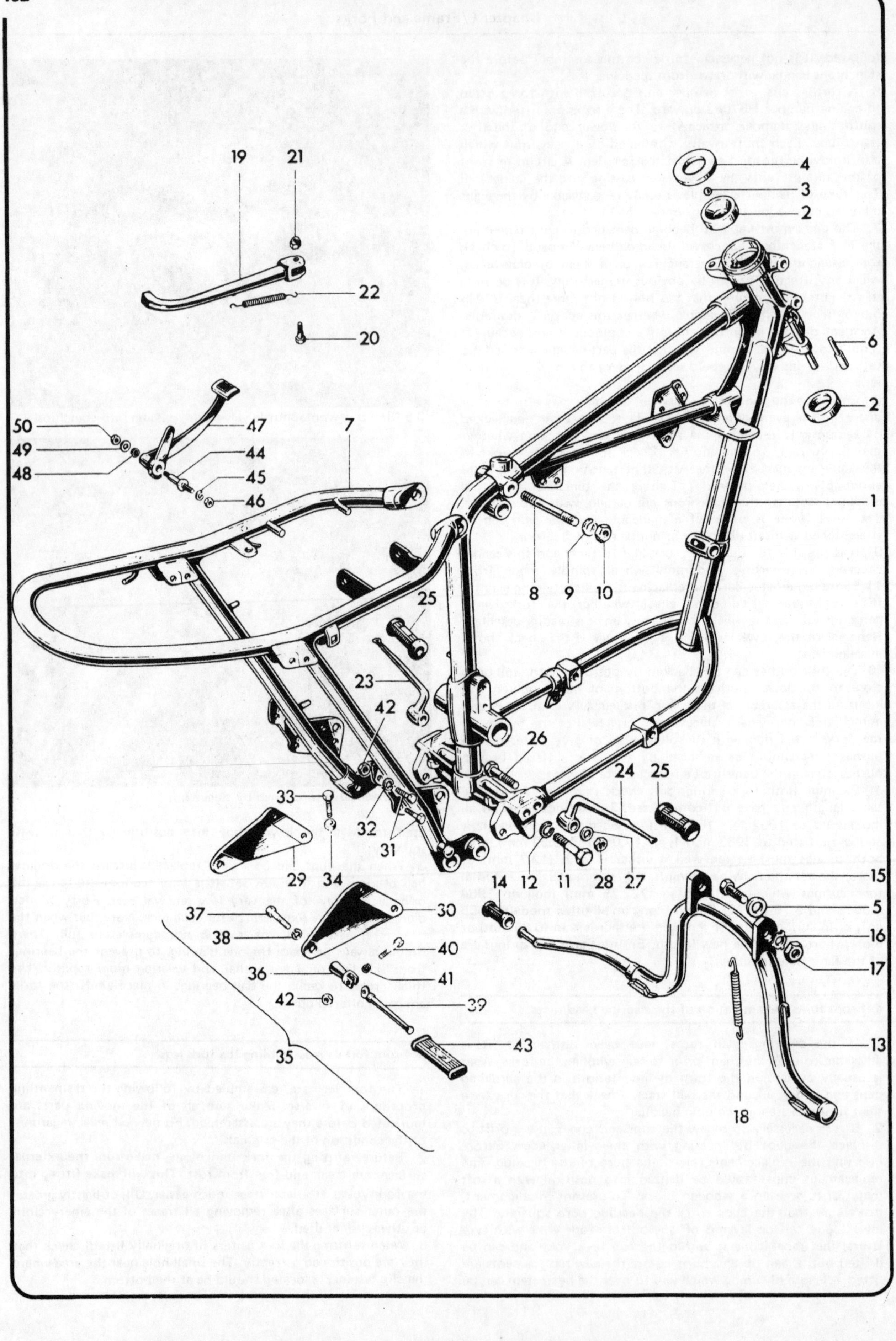
19
21
4
3
2
22
20
6
2
50
47
49
44
48
45
7
46
1
8
9
10
25
23
42
26
24
25
33
32
31
29
34
14
12
11
28
27
15
5
30
37
38
40
16
17
36
41
42
39
13
43
35
18

Fig. 6.3. Frame

1	Front frame 1 off		26	Bolt 2 off
2	Steering race cup 2 off		27	Plain washer 2 off
3	Steel ball (¼ in. dia.) 40 off		28	Nut 2 off
4	Top cone and dust cover 1 off		29	Left pillion footrest support 1 off
5	Washer 2 off		30	Right pillion footrest support 1 off
6	Steering stops 2 off		31	Bolt 4 off
7	Rear frame 1 off		32	Spring washer 4 off
8	Stud 1 off		33	Pedal stop screw 1 off
9	Spring washer 2 off		34	Nut 2 off
10	Nut 2 off		35	Pillion footrest 2 off
11	Bolt 2 off		36	Footrest hanger 2 off
12	Spring washer 2 off		37	Bolt 2 off
13	Centre stand 1 off		38	Spring washer 2 off
14	Pedal rubber 1 off		39	Pedal 2 off
15	Pivot bolt 2 off		40	Pivot bolt 2 off
16	Tab washer 2 off		41	Spring washer 2 off
17	Nut 2 off		42	Nut 6 off
18	Return spring 1 off		43	Pedal rubber 2 off
19	Prop stand 1 off		44	Brake pedal spindle 1 off
20	Pivot bolt 1 off		45	Spring washer 1 off
21	Self-locking nut 1 off		46	Nut 1 off
22	Return spring 1 off		47	Brake pedal 1 off
23	Left footrest 1 off		48	Spring washer 1 off
24	Right footrest 1 off		49	Plain washer 1 off
25	Pedal rubber 2 off		50	Nut 1 off

6 Front forks - refitting to frame

1 If it has been necessary to remove the fork assembly complete from the frame, refitting is accomplished by following the dismantling procedure in reverse. Check that none of the ball bearings are displaced whilst the steering head stem is passed through the headlug; it has been known for a displaced ball to fall into the headlug and wear a deep groove around the headstem of the lower fork yoke.

2 Take particular care when adjusting the steering head bearings. The blind or sleeve nut should be tightened sufficiently to remove all play from the steering head bearings and no more. Check for play by pulling and pushing on the fork ends and make sure the handlebars swing easily when given a light tap on one end.

3 It is possible to overtighten the steering head bearings and place a loading of several tons on them, whilst the handlebars appear to turn without difficulty. On the road, overtight head bearings cause the steering to develop a slow roll at low speeds.

4 Before the plated top fork nuts are replaced, do not omit to replace the drain plug in each fork leg and to refill each leg with the correct quantity and grade of oil.

5 Difficulty will be experienced in raising the fork stanchions so that their end taper engages with the taper inside the top fork yoke. Triumph service tool Z170 (unified threads) or Z161 (CEI threads) is specified for this purpose; if the service tool is not available, a wooden broom handle screwed into the inner threads of the fork stanchion can often be used to good effect.

6 Before final tightening, bounce the forks several times so that the various components will bed down into their normal working

6.5 Slot in brake plate must align with peg on fork leg

locations. This same procedure can be used if the forks are twisted, but not damaged, as the result of an accident. Always retighten working from the bottom upward.

7 Frame assembly - examination and renovation

1 If the machine is stripped for a complete overhaul, this affords a good opportunity to inspect the frame for cracks or other damage which may have occurred in service. Check the front down tube immediately below the steering head and the top tube immediately behind the steering head, the two points where fractures are most likely to occur. The straightness of the tubes

concerned will show whether the machine has been involved in a previous accident.

2 If the frame is broken or bent, professional attention is required. Repairs of this nature should be entrusted to a competent repair specialist, who will have available all the necessary jigs and mandrels to preserve correct alignment. Repair work of this nature can prove expensive and it is always worthwhile checking whether a good replacement frame of identical type can be obtained from a breaker or through the manufacturer's Service Exchange Scheme. The latter course of action is preferable because there is no safe means of assessing on the spot whether a secondhand frame is accident damaged too.

3 The part most likely to wear during service is the pivot and bush assembly of the swinging arm rear fork. Wear can be detected by pulling and pushing the fork sideways, when any play will immediately be evident because it is greatly magnified at the fork end. A worn pivot bearing will give the machine imprecise handling qualities which will be most noticeable when traversing uneven surfaces.

8 Swinging arm pivot bushes - examination and renovation

After engine number H49832

1 If play is felt from side to side, when clutching the rear wheel, it is probable that the swinging arm bushes need replacing.

2 Support the machine on the centre stand and remove the return spring off the stand.

3 Unscrew the rear brake adjusting rod nut and place it to one side.

4 Remove the rear chain spring link and remove the nut holding on the torque stay to the back brake plate.

5 Slacken the nut to the rear of the left hand rear suspension unit bottom mounting, and lift the rear of the chainguard.

6 Disconnect the speedometer cable. Remove the rear wheel as described in Chapter 7.

7 Release the clips holding the speedometer cable to the swinging arm and disconnect the rubber pipe from the chain oiler, which is attached to the torque stay.

8 Remove the complete exhaust system. Refer to Chapter 1.

9 Disconnect the brake light switch and the electrical snap connectors.

10 Take off the front footrests and the rear brake pedal.

11 Remove the small front chainguard and remove the lower nuts holding the oil tank to the frame. Note the position of the distance piece over the top stud, between the oil tank bottom bracket and the frame lug.

12 Remove the rear mudguard securing bolts and the mudguard.

13 Remove both swinging arm spindle nuts after bending back the tab washers; the tab washers come away with the bolt. (For reassembly, the thin-angled tabs locate into the sides of the frame.) The rear subframe can now be hinged out of the way and supported.

14 Take off the left and right hand distance pieces and note that the thicker of the two fits on the chain side, and that the right hand one is heavily ribbed. On reassembly, the ribbed side must abut against the rear frame side plate.

15 The spindle can now be drifted out with a soft drift.

16 To replace the bushes a good vice is needed. Having obtained the new bushes, use a tube, bigger than the outside diameter of the bush, and slightly longer and place it against the swinging arm. Line up the new bush with the old bush on the opposite side, so that when the vice is tightened, the new bush is pushed in, displacing the old one. Continue until the end of the bush is a flush fit. Repeat for the other side.

17 Line ream to a fit which just allows the swinging arm to drop under its own weight.

18 On reassembly, shimming for side play will probably be necessary until the side to side play is down to about 0.005 inch.

19 Reassemble in reverse order of dismantling and grease well.

20 Do not forget to bend over the tab washers.

Before engine number H49832

21 Unscrew the bolt at the front of the chainguard. Disconnect the stop lamp wires, remove the stop switch clip and the chainguard bolts. Remove the chainguard.

22 Take off the chain, speedometer cable and rear wheel.

23 Remove the two bottom rear suspension unit bolts.

24 To remove the spindle, first extract the retaining rods and then drift out. A threaded extractor can be used but it will have to be made, as there is no Triumph tool for this task.

25 When drifting out it is best to support the swinging arm at the opposite end.

26 Gather up the shims and spacing washers after pulling out the swinging arm.

27 Clean all parts with paraffin. Renew all worn parts and assemble in the reverse order.

9 Rear suspension units - examination

1 If the damping action is lost, renewal will be necessary because the dampers are sealed units. Always renew the damper units in pairs.

2 To inspect the springs and to check the damping action, take off each unit one at a time. Before releasing the springs, make sure that the unit is in its softest position. (Lowest position on three ramps available).

3 Grip the spring firmly, or if a shroud is fitted, grip this and press firmly down. When the two split collets in the top of the unit are free, flick them out with your thumb.

4 Wash the assembly thoroughly in paraffin and check for the damping action. It should be possible to push the damper rod inward quite rapidly but resistance should be felt when it is pulled out rapidly. The faster the action the harder should be the resistance. If no resistance is felt both damper units must be renewed.

5 To check the springs see the Specifications Section of this Chapter for the free length measurements.

6 If external shrouds are used, grease them inside with good grease prior to reassembly, to stop chafing.

7 Reassemble in reverse order. **Do not forget to attach and secure the rear brake torque arm.**

10 Rear suspension units - adjusting the loading

1 As mentioned earlier, the units can be adjusted without detaching them from the machine, so that the loading can be adjusted to suit the conditions under which the machine is to be used. A built-in cam at the lower end of each unit allows the sleeve carrying the lower end of the compression springs to be rotated, so that the sleeve is adjustable to different heights. The lowest position, or lightest loading, is recommended for solo riding, the middle or medium loading for a heavy solo rider or one with luggage, and the highest or heaviest loading when a pillion passenger is carried. A special C spanner, provided with the tool kit, is used for these adjustments.

2 Both units must always be set to an identical rating, otherwise the handling of the machine will be badly affected.

11 Removing and replacing the rear panels

1 Early models were fitted with panels enclosing the rear portion of the frame and it is necessary to remove these panels before any repair work can be undertaken.

2 To remove the panels, withdraw the four bolts which retain the dualseat hinges and the screw which holds the dualseat check wire. Lift the dualseat off the machine.

3 Remove the two screws at the front of the panels, facing the carburettor. Remove the rear petrol tank mounting bolt and unscrew the uppermost nuts of the pillion footrest plates. Remove the four screws in the top rear frame tube and the rear fixing bolt and nut, all of which are accessible from within the

9.2 Remove upper and lower bolts to free suspension units

9.7 Do not forget to fit and tighten rear brake torque arm

panels.

4 Disconnect the rear lamp wires at the snap connectors, then spring the panels apart and pull them backwards off the machine.

5 When replacing the panels, follow the dismantling procedure in reverse but do not tighten any of the fixing bolts and screws until all are in position.

12 Centre and prop stands - examination

1 The centre stand is attached to lugs on the bottom frame tubes to provide a convenient means of parking the machine or raising either wheel clear of the ground. The stand pivots on bolts through these lugs and is held retracted, when not in use, by a return spring.

2 The condition of the return spring and the return action of the stand should be checked regularly, also the security of the two nuts and bolts retained by the tab washer. If the stand drops whilst the machine is in motion, it may easily catch in the ground and cause the rider to come off.

3 Some riders remove the centre stand completely as it is inclined to ground if the machine is cornered vigorously. It is questionable whether such action is justifiable because in the

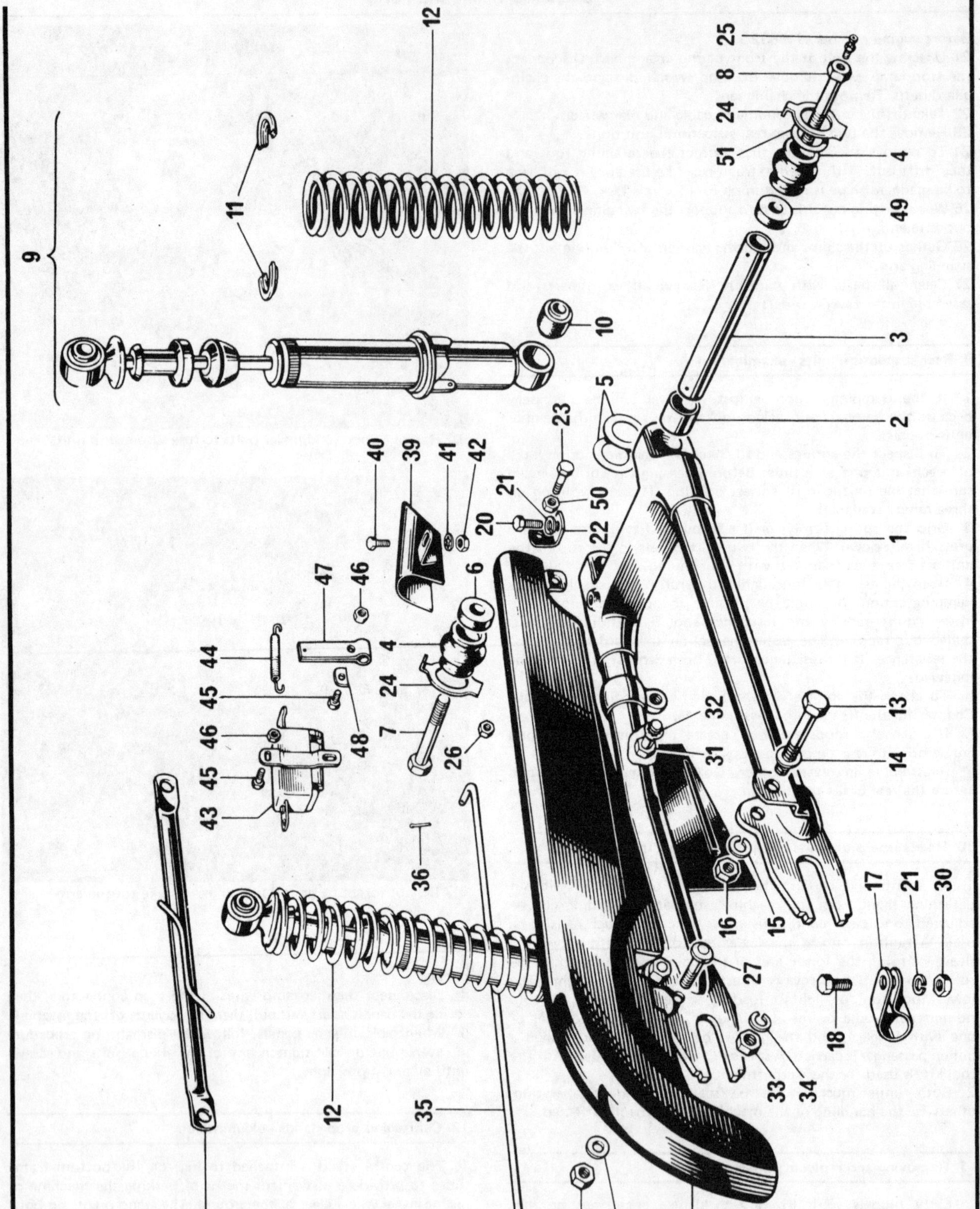

1
2
3
4
5
6
7
8
9
10
11
12
13
14
15
16
17
18
19
20
21
22
23
24
25
26
27
28
29
30
31
32
33
34
35
36
37
38
39
40
41
42
43
44
45
46
47
48
49
50
51

Fig. 6.4. Swinging arm fork and suspension units

1	Swinging fork 1 off	25	Lubricator 1 off
2	Pivot bush 2 off	26	Self-locking nut 1 off
3	Spindle 1 off	27	Stepped bolt 1 off
4	Gaiter 2 off	28	Plain washer 1 off
5	Shim 0.003 in.	29	Nut 1 off
5	Shim 0.005 in.	30	Nut UN 1 off
6	Left distance piece 1 off	31	Bolt 1 off
7	Left bolt 1 off	32	Serrated washer 1 off
8	Right bolt 1 off	33	Nut 1 off
9	Suspension unit (64052106) 2 off	34	Spring washer 1 off
9	Suspension unit (64052107) 2 off	35	Brake operating rod 1 off
10	Bonded bush (64533645) 4 off	36	Split pin 1 off
11	Spring retainer (9054/171) 4 off	37	Adjuster nut 1 off
12	Spring, 145 lbs chrome (64544234) 2 off	38	Brake torque stay 1 off
12	Spring, 100 lbs chrome (64543708) 2 off	39	Front chainguard 1 off
13	Unit fixing bolt 4 off	40	Bolt 1 off
14	Plain washer 2 off	41	Serrated washer 1 off
15	Spring washer 2 off	42	Nut UN 1 off
16	Nut 2 off	43	Stop lamp switch (54033234) 1 off
17	Clip 1 off	44	Spring (315738) 1 off
18	Bolt 1 off	45	Bolt 3 off
19	Chainguard 1 off	46	Self-locking nut 5 off
20	Bolt 1 off	47	Lever 1 off
21	Spring washer 2 off	48	"D" washer 1 off
22	Front bracket 1 off	49	Right distance piece 1 off
23	Bolt 1 off	50	Nut 1 off
24	Tab washer 2 off	51	Thrust washer 1 off

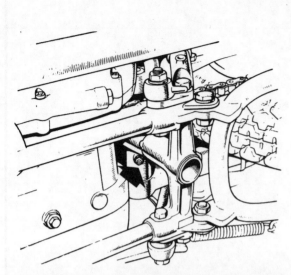

Fig. 6.5. Swinging arm fork lubrication nipple (prior to engine No. H 49833)

event of a puncture there is no other way of supporting the machine whilst either wheel is removed.

4   A prop stand which pivots from the left hand lower frame tube is provided for occasional parking, when it is not considered necessary to use the centre stand.

5   At regular intervals check that the prop stand return spring is in good order and that the pivot nut and bolt are not working loose.

13 Footrests and rear brake pedal - examination

1   The footrests are attached to the rear engine plates by a nut and washer behind each plate. They are prevented from rotating by two small pegs engaging with holes in the engine plates. Some of the earlier models have footrests which engage with mounting lugs below the engine unit.

2   If the machine is dropped, it is probable that the footrests will bend; they are quite soft. To straighten, they must be removed from the machine and their rubbers detached. They should be held in a vice during the straightening operation, using a blow lamp to heat the area where the bend occurs to a cherry red. If they are bent whilst cold there is risk of fracture.

3   The rear brake pedal passes through the rear engine plates where it engages with the lever carrying the brake rod. If the lever should bend, it can be straightened in a similar manner, after removal from the machine.

14 Oil tank - removal and replacement

1   Remove the oil filler cap, place a drip tray under the oil tank, remove the drain plug and allow the tank to drain.

2   Unscrew the flange nut on the oil feed and undo the clips on the remaining oil pipes.

3   On machines prior to H57083, remove the tool tray.

4   Remove the battery and carrier.

5   Remove the battery carrier frame cross strap by detaching the bolts at either end, noting the washer positions. Disconnect and remove the two switch panel nuts on machines with engine numbers prior to H57083.

6   Withdraw each battery carrier frame cross strap.

7   Remove the top fixing brackets. Disconnect the dualseat check wire and disengage the screwed studs clear of the frame brackets by pushing them through and out of the spigot rubber bushes in the oil tank top brackets.

8   Swing the oil tank in towards the space above the gearbox,

and lift the oil tank off the bottom spigot.

9   If difficulty is encountered, remove the left hand panel by undoing the knurled plastic nut at the top.

10   Unscrew the two bottom mounting bolts and remove the bracket.

11   Wash out the tank and inspect it for leaks. Wash out the filter or feed pipe and replace the fibre washer, then reassemble in the reverse order of stripping.

12   **Do not forget to refill with clean oil. Tighten all pipe unions very securely.**

15 Speedometer head and tachometer head - removal and replacement

1   The speedometer and tachometer heads are each secured to a bracket that bolts to the fork top yoke. The bracket is rubber-mounted to damp out vibration; both instruments are held to the bracket by studs which project from the base of the casing. On late models the instruments are mounted within rubber cups, attached to brackets that fit under each top fork nut.

2   To remove either instrument, detach the drive cable by unscrewing the gland nut where the cable enters the instrument body. Pull out the bulb holder complete with the bulb used to illuminate the dial and unscrew the two nuts that secure the instrument to the mounting bracket, taking care not to displace the shakeproof washers. The instrument can now be lifted away.

3   Apart from defects in either the drive or the drive cable, a speedometer or tachometer that malfunctions is difficult to repair. Fit a replacement, or alternatively entrust the repair to an instrument repair specialist, bearing in mind that an efficient speedometer is a statutory requirement. If, in the case of a speedometer, the mileage recordings also cease, it is highly probable that either the drive cable or the drive is at fault and not the speedometer head itself. It is very rare for all recordings to fail simultaneously.

16 Speedometer and tachometer drive cables - examination and renovation

1   It is advisable to detach the speedometer and tachometer drive cables from time to time, in order to check whether they are adequately lubricated and whether the outer covers are compressed or damaged at any point along their run. A jerky or sluggish movement at the instrument head can often be attributed to a cable fault.

2   To grease the cable, uncouple both ends and withdraw the inner cable. After removing the old grease, clean with a petrol soaked rag and examine the cable for broken strands or other damage.

3   Regrease the cable with high melting point grease, taking care not to grease the last six inches closest to the instrument head. If this precaution is not observed, grease will work into the instrument and immobilise the sensitive movement.

4   If the cable breaks, it is usually possible to renew the inner cable alone, provided the outer cable is not damaged or compressed at any point along its run. Before inserting the new inner cable, it should be greased in accordance with the instructions given in the preceding paragraph. Try and avoid tight bends in the run of a cable because this will accelerate wear and make the instrument movement sluggish.

17 Dualseat - removal

1   The dual seat hinges on the left hand side of the machine when the retaining catch on the right hand side, below the dual seat, is released. A restraining wire limits the amount of travel.

2   To remove the dualseat, detach the restraining wire from the seat underpan where it is retained by a self-tapping screw. If either the forward or rearward hinge is removed, by unscrewing

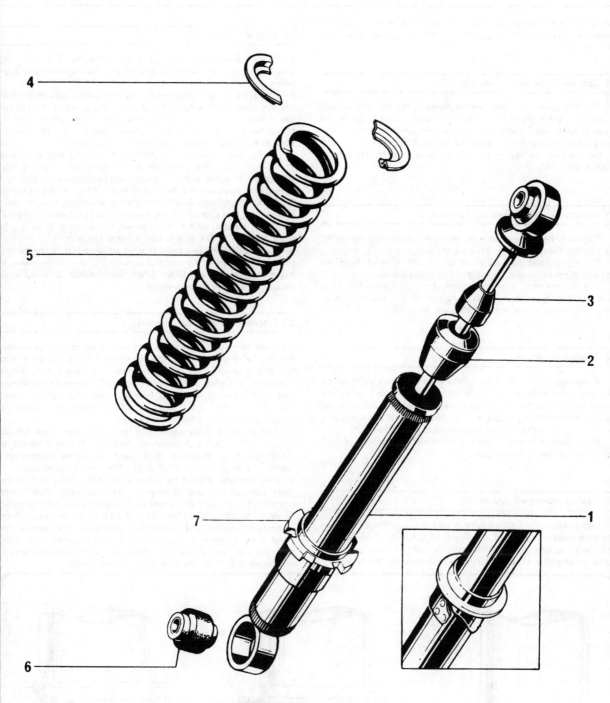

Fig. 6.6. Rear suspension unit
(Inset shows earlier type of adjuster)

1   Damper unit
2   Bump stop
3   Bump stop
4   Spring retainers
5   Coiled spring
6   Bonded bush
7   Static lead adjuster

the two bolts that secure it to the tapped inserts in the seat underpan, the seat can be slid off the remaining hinge pivot and lifted away.

18 Left hand side cover - removal

1   The left hand side cover, which contains the tool roll, is held in position by a panel knob having a large milled head. If the knob is unscrewed completely, the cover can be drawn off the two projections from the rear subframe and lifted away.
2   Do not lose the two grommets which fit over the frame tube projections.

19 Petrol tank embellishments - removal

1   The tank badges are held on with two crosshead screws in each badge. Remove them to release the badge.
2   The leg or knee pads are glued on like the soles of shoes; do not pull them off unless you intend to replace them.
3   The chrome strip down the centre of the tank is held on by a hook at the front and is held captive under the rear tank mounting. Remove the nut and lift away the strip.

20 Steering head lock

1   A steering head lock is built into the upper fork yoke which will lock the handlebars on full left lock when it is operated. The lock is held in position by a grub screw blanked off with a sealing washer. In operation, a tongue from the lock projects through a plate attached to the steering head when aligned correctly.
2   If the lock is changed, the key must also be changed, as the locks are numbered individually.

21 Fairing attachment points

Some models are fitted with a cast-in lug at the top of the steering head to facilitate the fixing of a fairing. Two rod-like attachments pass through the base of the steering head, the lower extremeties of which form the lock stops for the forks.

22 Sidecar alignment

1   Using conventional fittings, little difficulty is experienced when attaching a sidecar to any of the Triumph construction twins.
2   Good handling characteristics of the outfit will depend on the accuracy with which the sidecar is aligned. Provided the toe-in and lean-out are within prescribed limits, good handling characteristics should result, leaving scope for other minor adjustments about which opinions vary quite widely.
3   To set the toe-in, check that the front and rear wheels of the motor cycle are correctly in line and adjust the sidecar fittings so that the sidecar wheel is approximately parallel to a line drawn between the front and rear wheels of the machine. Re-adjust the fittings so that the sidecar wheel has a slight toe-in toward the front wheel of the motor cycle, as shown in Fig 6.8. When the amount of toe-in is correct, the distance 'B' should be from 3/8 inch to 3/4 inch less than the distance at 'A'.
4   Lean-out is checked by attaching a plumb line to the handlebars and measuring the distance between 'C' and 'D' as shown in Fig 6.8. Lean-out is correct when the distance 'C' is approximately 1 inch greater than at 'D'.

23 Cleaning the machine - general

1   After removing all surface dirt with a rag or sponge washed frequently in clean water, the application of car polish or wax will give a good finish to the machine. The plated parts should require only a wipe over with a damp rag, followed by polishing with a dry rag. If, however, corrosion has taken place, which may occur when the roads are salted during the winter, a proprietary chrome cleaner can be used.
2   The polished alloy parts will lose their sheen and oxidise slowly if they are not polished regularly. The sparing use of metal polish or a special polish such as Solvol Autosol will restore the original finish with only a few minutes labour.
3   The machine should be wiped over immediately after it has been used in the wet so that it is not garaged under damp conditions which will encourage rusting and corrosion. Make sure to wipe the chain and if necessary re-oil it to prevent water from entering the rollers and causing harshness with an accompanying rapid rate of wear. Remember there is little chance of water entering the control cables if they are lubricated regularly, as recommended in the Routine Maintenance Section.

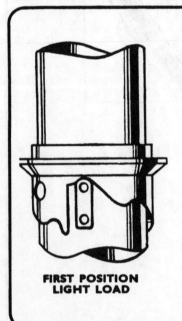 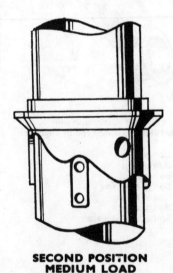 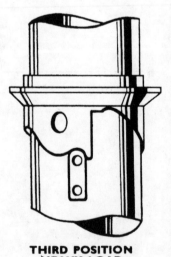

FIRST POSITION   **SECOND POSITION**   **THIRD POSITION**
LIGHT LOAD   **MEDIUM LOAD**   **HEAVY LOAD**

Fig.6.7. Adjusting the rear suspension units

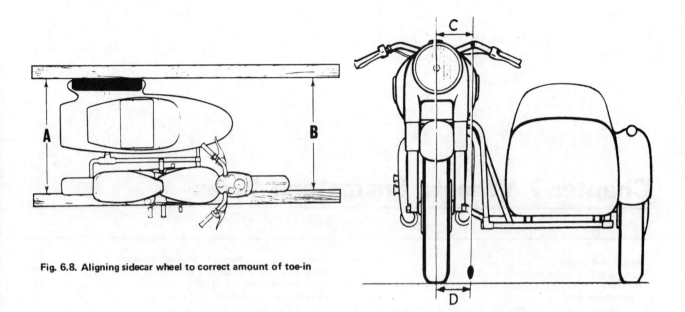

Fig. 6.8. Aligning sidecar wheel to correct amount of toe-in

Fig. 6.9. Setting the amount of lean-out, using a plumb line

24 Fault diagnosis

Symptom	Reason/s	Remedy
Machine is unduly sensitive to road surface irregularities	Fork and/or rear suspension units damping ineffective	Check oil level in forks. Renew suspension units.
Machine rolls at low speeds	Steering head bearings overtight or damaged	Slacken bearing adjustment. If no improvement, dismantle and inspect head races.
Machine tends to wander. Steering imprecise	Worn swinging arm suspension bearings	Check and if necessary renew bushes.
Fork action stiff	Fork legs twisted in yokes or bent	Slacken off wheel spindle clamps, yoke pinch bolts and fork top nuts. Pump forks several times before retightening from bottom. Straighten or renew bent forks.
Forks judder when front brake is applied	Worn fork bushes Steering head bearings slack	Strip forks and renew bushes. Re-adjust to take up play.
Wheels seem out of alignment	Frame distorted through accident damage	Check frame after stripping out. If bent, specialist repair or renewal is necessary.
Machine handles badly under all types of condition	General frame and fork distortion as result of a previous accident or broken frame tube	Strip frame and check for alignment very carefully. If wheel track is out, renew or straighten parts involved. Broken tube will be self-evident.

Chapter 7 Wheels, Brakes and Tyres

Contents

Specifications

Model T100R — Daytona

Wheels

Rim size - front/rear	WM2-19/WM3-18
Type:	
Front	Spoke — Single cross lacing
Rear	Spoke — Double cross lacing
Spoke details - front:	
Left side	20 off 10 SWG $5^5/8$ in UK (142.86 mm)
Right side	20 off 10 SWG $4^{11}/16$ in UH (116.1 mm)
Spoke details - rear:	
Left side	20 off 10 SWG $7^9/16$ in UH (102.1 mm)
Right side	20 off 10 SWG $7^7/8$ in UH (199 mm)

Wheel bearings

Front and rear, dimensions and type	20 x 47 x 14 mm Ball journal
Front spindle diameter (at bearing journals)	0.7868 - 0.7873 in (19.98 - 19.99 mm)
Rear spindle diameter (at bearing journals)	0.7862 - 0.7867 in (19.97 - 19.98 mm)

Q.D. rear wheel

Bearing type	$3/4$ x $1^7/8$ x $9/16$ in Ball journal (19 x 47.6 x 14.3 mm)
Bearing sleeve - journal diameter	0.7490 - 0.7495 in (18.92 - 18.94 mm)
Brake drum bearing	$7/8$ x 2 x $9/16$ in (22.23 x 50.8 x 14.3 mm)
Bearing sleeve - journal diameter	0.8740 - 0.8745 in (22.199 - 22.21 mm)
Bearing housing - internal diameter	1.9980 - 1.9990 in (25.35 - 25.37 mm)

Rear wheel drive

Gearbox sprocket	See 'Gearbox'
Rear wheel sprocket teeth	46
Chain details:	
No of links - solo	102
Pitch	$5/8$ in (15.87 mm)
Width	$3/8$ in (9.52 mm)
Speedometer gearbox drive ratio	19 : 10

Brakes

Type:

    Front   ... ...   Internal expanding, twin leading shoe

    Rear   ... ...   Internal expanding

Diameter - front and rear   ... ...   7 in (177.8 mm)

Lining thickness ... ...   0.179 - 0.190 in (4.54 - 4.83 mm)

Tyres

Size:

    Front   ... ...   3.25 x 19 in Dunlop K70

    Rear   ... ...   4.00 x 18 in Dunlop K70

Tyre pressure:

    Front   ... ...   24 lb sq in (1.7 kg sq cm)

    Rear   ... ...   25 lb sq in (1.7 kg sq cm)

Model T100C — Trophy 500

Brakes

Type - front   ... ...   Internal expanding twin leading show 7 in (177.8 mm)

Tyres (East Coast models only)

Front   ... ...   3.25 x 19 Trials Universal

Rear   ... ...   4.00 x 19 Trials Universal

Rear chain   ... ...   102 pitches

Model T100S — Tiger 100

Wheels

Rim size - front and rear WM2-18

Type:

    Front   ... ...   Spoke — Single cross lacing

    Rear   ... ...   Spoke — Double cross lacing

Spoke details:

    Front   ... ...   40 off 8/10 SWG butted $5^{17}/32$ in UH straight

    Rear - left side   ... ...   20 off 8/10 SWG butted $7^{9}/16$ in UH 90°

       - right side   ... ...   20 off 8/10 SWG butted $7^{7}/8$ in UH 90°

Brakes

Type ... ...   Internal expanding

Drum diameter:

    Front   ... ...   7 in   (177.8 + 0.0509 mm)

    Rear   ... ...   7 in + 0.002 in (177.8 + 0.0508 mm)

Lining thickness - front and rear ... ...   0.187 - 0.197 in (4.75 - 5.0 mm)

Lining area - front and rear   ... ...   14.6 sq in (93.7 cm^2)

Tyres

Size:

    Front   ... ...   3.25 x 18 in Dunlop ribbed (82.5 x 457.2 mm)

    Rear   ... ...   3.50 x 18 in Dunlop K70 (88.9 x 457.2 mm)

               (USA T100C, T100R 4.0018 in) (101.6 x 457.2 mm)

Tyre pressure:

    Front   ... ...   24 lb sq in (1.7 kg sq cm)

    Rear   ... ...   25 lb sq in (1.7 kg sq cm)

Model T100T — Daytona Sports

Brakes

Drum diameter - front ... ...   8 in (203.2 mm)

Lining thickness - front   ... ...   0.183 in (4.65 mm)

Lining area - front   ... ...   23.4 sq in (150.9 cm^2)

Rear wheel drive

Rear chain - no of links...   102

Model T90 — Tiger 90

Wheels

Rim size:

    Front WM2-18

    Rear WM2-18

Spoke details - front   ... ...   40 off 8/10 SWG butted $5^{17}/32$ in UH straight

Rear wheel drive

Gearbox sprocket teeth	See 'Gearbox'
Rear wheel sprocket teeth	46
Chain details:	
No of links	102
Pitch	5/8 in (15.87 mm)
Width	3/8 in (9.52 mm)
Speedometer gearbox drive ratio	19 : 10

Tyres

Size:	
Front	3.25 x 18 in
Rear	3.50 x 18 in
Tyre pressures:	
Front	24 lb sq in (1.7 kg sq cm)
Rear	24 lb sq in (1.7 kg sq cm)

Model 5TA — Speed Twin
Wheels

Rim size:	
Front	WM2-18
Rear	WM2-18
Spoke details - front	40 off 8/10 SWG butted 5^{17}/32 in UH straight

Rear wheel drive

Gearbox sprocket teeth	See 'Gearbox'
Rear wheel sprocket teeth	46
Chain details:	
No of links	103
Pitch	5/8 in (15.87 mm)
Width	3/8 in (9.53 mm)
Speedometer gearbox drive ratio	2 : 1

Tyres

Size:	
Front	3.25 x 18 in Avon Speedmaster
Rear	3.50 x 18 in Avon Speedmaster
Tyre pressures:	
Front	24 lb sq in (1.7 kg sq cm)
Rear	24 lb sq in (1.7 kg sq cm)

Model 3TA — Twenty One
Wheels

Rim size:	
Front	WM2-18
Rear	WM2-18
Spoke details - front	40 off 8/10 SWG butted 5^{17}/32 in UH straight

Rear wheel drive

Gearbox sprocket teeth	17
Rear wheel sprocket teeth	46
Chain details:	
No of links	102
Pitch	5/8 in (15.87 mm)
Width	3/8 in (9.53 mm)
Speedometer gearbox drive ratio	2 : 1

Tyres

Size:	
Front	3.25 x 18 in
Rear	3.50 x 18 in
Tyre pressures:	
Front	24 lb sq in (1.7 kg sq cm)
Rear	24 lb sq in (1.7 kg sq cm)

Model 3TA — Twenty One
Sports models fitted with AC magneto (engine numbers prior to H 57083)

Front wheel

Rim size	WM2-19
Type	Spoke — Single cross lacing
Spoke details	40 off 8/10 SWG butted 6 in UH straight
Unless otherwise stated, T100R data applies to all model.	

1 General description

Since the inception of the 500 cc Triumph twins, a variety of wheel sizes has been used. The latest models have a WM2 - 19 in front wheel and a WM2 - 18 in rear wheel, with the exception of the export Trophy 500. This latter model has both wheels of 19 in diameter.

Early models have a single leading shoe front brake, but as performance gradually increased as the result of continuing engine modifications, it has been found necessary to employ a twin leading shoe front brake on the later models. All models, irrespective of age, have a single leading shoe rear brake.

Some models have a quickly-detachable rear wheel, which enables the wheel to be removed independently of the rear sprocket and chain. This greatly facilitates the repair of punctures.

2 Front wheel - examination and renovation

1 Place the machine on the centre stand so that the front wheel is raised clear of the ground. Spin the wheel and check for rim alignment. Small irregularities can be corrected by tightening the spokes in the affected area, although a certain amount of skill is necessary if over correction is to be avoided. Any 'flats' in the wheel rim should be evident at the same time. These are more difficult to remove with any success and in most cases the wheel will have to be rebuilt on a new rim. Apart from the effect on stability, there is greater risk of damage to the tyre bead and walls if the machine is run with a deformed wheel, especially at high speeds.

2 Check for loose or broken spokes. Tapping the spokes is the best guide to the correctness of tension. A loose spoke will produce a quite different note and should be tightened by turning the nipple in an anticlockwise direction. Always check for run-out by spinning the wheel again.

3 If several spokes require retensioning or there is one that is particularly loose, it is advisable to remove the tyre and tube so that the end of each spoke that projects through the nipple after retensioning can be ground off. If this precaution is not taken, the portion of the spoke that projects may chafe the inner tube and cause a puncture.

3 Front drum brake assembly - examination, renovation and reassembly

1 The front brake assembly complete with brake plate can be withdrawn from the front wheel by following the procedure in Chapter 6, Section 2, paragraphs 1 to 3.

2 An anchor plate nut retains the brake plate on the front wheel spindle. When this nut is removed, the brake plate can be drawn away, complete with the brake shoe assembly.

3 Examine the condition of the brake linings. If they are wearing thin or unevenly, the brake shoes should be relined or renewed.

4 To remove the brake shoes from the brake plate, pull them apart whilst lifting them upward, in the form of a V. When they are clear of the brake plate, the return springs can be removed and the shoes separated. Do not lose the abutment pads fitted to the leading edge of each shoe.

5 The brake linings are rivetted to the brake shoes and it is easy to remove the old linings by cutting away the soft metal rivets. If the correct Triumph replacements are purchased, the new linings will be supplied ready-drilled with the correct complement of rivets. Keep the lining tight against the shoe throughout the rivetting operation and make sure the rivets are

3.2 Front brake plate will lift off hub

3.4 Pull and lift brake shoes upwards to separate from brake plate

countersunk well below the lining surface. If workshop facilities and experience suggest it would be preferable to obtain replacement shoes, ready lined, costs can be reduced by making use of the Triumph service exchange scheme, available through Triumph agents.

6 Before replacing the brake shoes, check that both brake operating cams are working smoothly and not binding in their pivots. The cams can be removed for cleaning and greasing by unscrewing the nut on each brake operating arm and drawing the arm off, after its position relative to the cam spindle has been marked so that it is replaced in exactly the same position. The spindle and cam can then be pressed out of the housing in the back of the brake plate.

7 Check the inner surface of the brake drum on which the brake shoes bear. The surface should be smooth and free from score marks or indentations, otherwise reduced braking efficiency is inevitable. Remove all traces of brake lining dust and wipe both the brake drum surface and the brake shoes with a clean rag soaked in petrol, to remove any traces of grease. Check that the brake shoes have chamfered ends to prevent pick-up or grab. Check that the brake shoe return springs are in good order and have not weakened.

8 To reassemble the brake shoes on the brake plate, fit the return springs first and force the shoes apart, holding them in a V formation. If they are now located with the operating cams they can usually be snapped into position by pressing downward. Do not use excessive force or the shoes may distort permanently. Make sure the abutment pads are not omitted.

4 Front wheel bearings - removal, examination and replacement

1   When the brake plate is removed, the bearing retainer within the brake drum will be exposed. This has a left-hand thread and is removed by using either Triumph service tool 61-3694 or a centre punch. The bearing on the right hand side can then be displaced by striking the left hand end of the front wheel spindle, using the shoulder of the spindle to drive the bearing outward. The left hand bearing can be displaced in similar fashion, after the retaining circlip has been removed, if the wheel spindle is inserted from the other side. Take note of the location of the inner and outer grease retaining plates (where fitted) so that they are replaced correctly during reassembly.
2   Wash the bearings in a petrol/paraffin mix to remove all traces of old grease and oil. Clean out the hub and repack it with fresh high melting point grease.
3   When the bearings are dry, check them for play or signs of roughness when they are turned. If there is any doubt about their condition, renew them.
4   When fitting the bearings, first insert the inner left hand grease retainer. Pack the left hand bearing with grease and drive it into place with a drift of the correct diameter. Fit the outer dust cap, followed by the circlip, which must locate correctly with the retaining groove. Insert the wheel spindle so that the shouldered end bears against the bearing from the inside of the hub and drive the bearing and grease retainer forward so that they are hard against the circlip holding them in position.
5   Withdraw the spindle and re-insert it the other way round. Refit the right hand grease retainer, repack the right hand bearing with grease and drive it into position. Screw in the retainer ring, remembering it has a left hand thread, and tighten it fully. Tap the spindle from the left hand end to centralise.

5 Front wheel - replacement

1   Place the front brake assembly in the brake drum and align the front wheel so that the torque anchorage locates with the peg or stud on the lower right hand fork leg. This is most important because the anchorage of the front brake plate is dependent solely on the correct location of these parts.
2   Whilst holding the wheel in position, replace the split clamps which secure the wheel spindle to each fork end. Each clamp has two bolts which must be tightened evenly and fully. Note that each end of the spindle has a groove which must locate with the clamp bolts or studs.

3   Reconnect the front brake cable and check that the brake functions correctly, especially if the adjustment has been altered or the brake operating arms have been removed and replaced during the dismantling operation. Recheck the tightness of the bolts in the split clamps.

6 Rear wheel - removal and examination

1   Place the machine on the centre stand and before removing the wheel, check for rim alignment, loose or broken spokes and other wheel defects by following the procedure applying to the front wheel, as described in Section 2.
2   Two types of rear wheel have been fitted, the standard or the quickly detachable type. The latter has the advantage of simplified removal, leaving the final drive chain and sprocket in position.
3   If the wheel is of the standard type, commence by disconnecting the final drive chain at the detachable spring link. The task is made easier if the link is first positioned so that it is on the rear wheel sprocket. Unwind the chain off the rear sprocket and lay it on a clean surface.
4   Take off the brake rod adjuster and pull the brake rod clear of the brake operating arm. Disconnect the torque stay by removing the nut where the stay joins the rear brake plate. If the stop lamp stays in the 'on' position, disconnect the snap connector in the lead.
5   Slacken both rear wheel spindle nuts and raise the rear chainguard by slackening the bottom nut on the left hand suspension unit. Remove the speedometer cable from the gearbox on the right hand side of the rear hub by unscrewing the gland nut and withdrawing the cable. Withdraw the wheel rearward until it drops from the frame ends complete with chain adjusters. It may be necessary to tilt the machine or raise it higher, so that there is sufficient clearance for the wheel to be taken away from the machine.
6   A different procedure is employed in the case of machines fitted with the quickly detachable rear wheel. It is necessary only to unscrew the gland nut from the speedometer drive gearbox and withdraw the cable, then unscrew and remove the wheel spindle from the right hand side of the machine. If the shouldered distance piece between the frame end and the hub is removed, the wheel can be pulled sideways to disengage it from the brake drum centre (splined fitting) before it is lifted away. Note there is a rubber ring seal over the splines which is compressed when the wheel is in position. This acts as a grit seal and must be maintained in a good condition.

4.1 Front wheel bearing retainer has left-hand thread

4.1a Left-hand bearing and dust cover are retained by circlip

7 Rear wheel bearings - removal, examination and replacement

1 On early models fitted with standard wheels remove both spindle nuts and the chain adjusters, then unscrew the brake anchor plate nut and withdraw the brake plate assembly and the spacer behind it. From the right-hand end remove the nut and washer, followed by the speedometer drive gearbox and its spacer, then unscrew the three bolts and withdraw the speedometer drive adaptor. Tap out the spindle, taking care not to damage its threaded ends. From the brake drum side remove the locking grub screw and use a peg spanner or a punch to unscrew the bearing retaining ring. **Note:** *check carefully whether the ring has a left- or right-hand thread before attempting to unscrew it and never use excessive force or the hub may be damaged.* Pass a drift through the hub from the opposite side to tap out each bearing in turn and withdraw the central spacer and bearing backing ring.

2 On reassembly place the backing ring against the hub shoulder on the brake drum side and tap in the left-hand bearing, then tighten the retaining ring securely and lock it with the grub screw. Turn the wheel over, pack the hub no more than 2/3 full with grease then refit the central spacer and the second bearing. The remainder of the reassembly procedure is a reversal of the dismantling sequence, but note that the speedometer gearbox drive dogs must engage with the adaptor slots and that the anchor plate spacer is refitted with its shouldered side outwards.

3 On late models fitted with standard wheels, remove the left-hand spindle nut and chain adjuster, then unscrew the retaining nut and withdraw the brake anchor plate assembly and spacer. Pull the spindle, with the speedometer drive gearbox and spacer, out to the right. From the brake drum side, remove the locking grub screw and use service tool 61-3694 or a peg spanner or punch to unscrew the bearing retaining ring. **Note:** *check carefully whether the ring has a left- or right-hand thread before attempting to unscrew it and never use excessive force or the hub may be damaged.* Insert a drift from the left-hand side and displace the central spacer so that a drift can be applied to the left-hand bearing inner race; it may be necessary to collapse the grease retainer. Pass a suitable drift through the hub from the right and tap out the left-hand bearing, then withdraw the backing ring, grease retainer and the central spacer. **Note:** *before attempting to remove the right-hand bearing, check whether it is retained by the speedometer drive adaptor; if this is the case the adaptor must be unscrewed before the bearing can be tapped out and the inner and outer grease retainers removed.* **Before** *unscrewing the adaptor, check carefully whether it has a left- or right-hand thread and do not use excessive force or the hub may be damaged.*

4 On reassembly, the left-hand bearing grease retainer can be carefully hammered flat to restore its original shape if it was damaged. Fit the right-hand bearing inner grease retainer, pack the bearing with grease and tap it into the hub, followed by the outer grease retainer. If removed, refit the speedometer drive adaptor and tighten it securely. Turn the wheel over, fit the central spacer, the grease retainer and the backing ring, then pack it with grease and tap in the left-hand bearing. Refit the retaining ring, tightening it securely, and lock it with the grub screw. The remainder of the reassembly procedure is a reversal of the dismantling sequence, but note that the speedometer gearbox drive dogs must engage with the adaptor slots and that the anchor plate spacer is refitted with its shouldered side outwards.

5 On early models fitted with quickly-detachable rear wheels, unscrew the locknut on the bearing sleeve right-hand end, then withdraw the washer, speedometer drive gearbox and its spacer. Unscrew the adjuster nut and remove the second spacer, then tap out to the left the bearing sleeve and withdraw the dust cover and both bearing inner races. The outer races can be tapped out by passing a drift through the hub from the opposite side; note the backing ring behind each.

6 On reassembly, place the backing rings against the hub shoulders and press in both outer races. Pack it with grease and fit the left-hand bearing inner race then press in the dust cover. Turn the wheel over and pack the hub no more than 2/3 full with grease, then fit the right-hand bearing inner race having packed it with grease. Fit the bearing sleeve from the left-hand (splined) side of the hub, then refit one of the spacers and the (thicker) adjusting nut. Tighten the nut hard to settle the bearings then slacken it by at least one flat (1/6 turn) until the sleeve is free to rotate easily. Fit the second spacer and the speedometer drive gearbox, aligning its drive dogs with the slots in the hub adaptor, then refit the washer and the (thin) locknut. Tighten the locknut securely then check the bearing adjustment; the sleeve should be able to rotate freely with the barest minimum of free play. If necessary, adjust the bearings until the setting is correct.

7 Later models (from 1965 on) fitted with quickly-detachable rear wheels use ball journal bearings instead of the previous taper roller types. Hold the bearing sleeve by the slot in its tapered end and unscrew the nut from its right-hand end, then withdraw the washer, speedometer drive gearbox and the spacer. Push the bearing sleeve out to the left and withdraw the left-hand bearing dust cap. **Note:** *before attempting to remove the right-hand bearing, check whether it is retained by the speedometer drive adaptor; if this is the case the adaptor must be unscrewed before the bearing can be tapped out.* **Before** *unscrewing the adaptor, check carefully whether it has a left- or right-hand thread and do not use excessive force or the hub may be damaged.* To remove the bearing itself, displace the central spacer (inner distance piece) as described in paragraph 3 above then drive it out with its outer grease retainer, using a drift passed through the hub from the opposite side. Withdraw the backing ring, inner grease retainer and central spacer (inner distance piece), then tap out the left-hand bearing and its inner grease retainer.

8 On reassembly, the right-hand bearing grease retainer can be carefully hammered flat to restore its original shape if it was damaged. Fit its inner grease retainer, then pack it with grease and tap the left-hand bearing into the hub, followed by its outer grease retainer. Turn the wheel over and fit the central spacer then fit the right-hand bearing in the same way. Refit and tighten securely the speedometer drive adaptor, if it was removed. The remainder of reassembly is a reversal of the removal procedure, but note that the speedometer gearbox drive dogs must engage with the slots in the adaptor.

9 On all models, check the bearings for wear as described in Section 4 and renew any worn or damaged components. Ensure the bearings are packed with high melting-point grease before they are refitted.

6.6 Unscrew gland nut to release speedometer cable

6.6a Quickly-detachable wheels have a pull-out spindle

6.6b Toothed arrangement locates wheel with brake drum and sprocket

8.2 Remove circlip and oil seal for access to bearing

Fig. 7.1. Front wheel, models T100T and T100R

1	Front wheel complete 1 off
2	Rim, spokes and hub assembly 1 off
3	Front wheel rim (WM2-19) 1 off
4	Spoke (8/10G x 5 5/8 in. straight) c/w nipple 20 off
5	Spoke (8/10G x 4 25/32 in. 78°) c/w nipple 10 off
6	Spoke (8/10G x 4 7/8 in. 100°) c/w nipple 10 off
7	Spoke nipple 40 off
8	Hub and brake drum 1 off
9	Wheel spindle 1 off
10	Bearing 2 off
11	Dust cover 1 off
12	Circlip 1 off
13	Bearing support ring 1 off
14	Grease retainer 1 off
15	Retaining ring 1 off
16	Cover plate 1 off
17	Balance weight, ½ oz.
18	Balance weight, 1 oz.
19	Brake anchor plate 1 off
20	Brake shoe c/w lining 2 off

21	Brake lining 2 off
22	Brass rivet 16 off
23	Abutment pad 2 off
24	Shoe return spring 2 off
25	Brake cam 2 off
26	Brake cam lever (front) 1 off
27	Nut 2 off
28	Brake cam lever (rear) 1 off
29	Rod assembly 1 off
30	Fork end, threaded 1 off
31	Locknut 1 off
32	Lever return spring 1 off
33	Anchor plate nut 1 off
34	Pivot pin 2 off
35	Split pin 1/16 in. diam. x 3/8 in. long 2 off
36	Plain washer 2 off
37	Anchor plate gauze (for air scoop) 1 off
38	Taptite screw 3 off
39	Washer for taptite screws 3 off

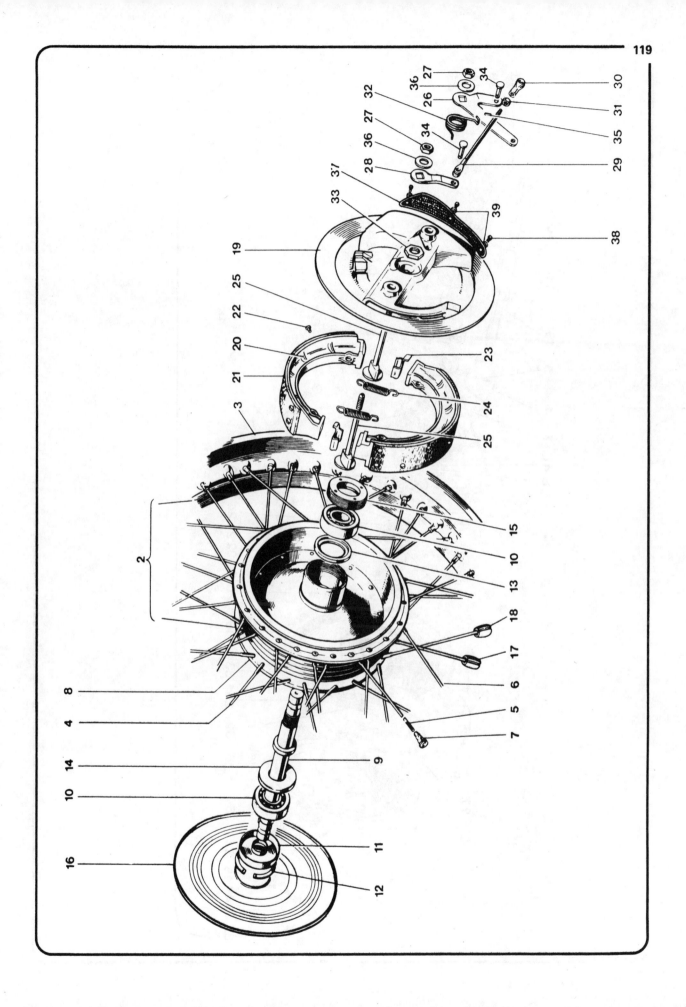

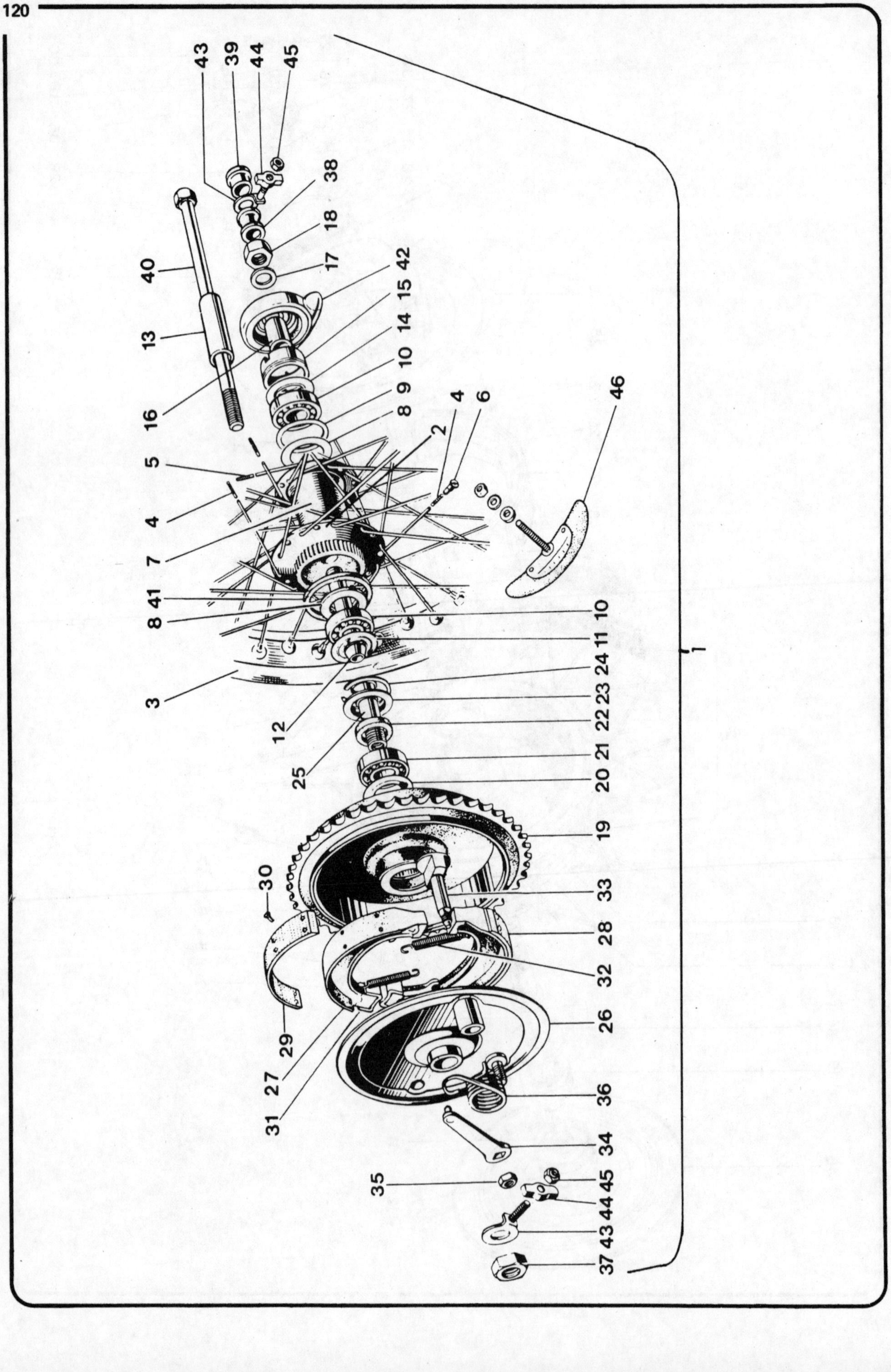
43
39
44
45
38
18
17
42
15
14
10
9
8
2
4
6
46
40
13
16
5
4
7
41
8
10
11
24
23
22
21
20
1
3
12
25
19
30
33
28
32
26
29
27
31
36
34
35
45
44
43
37

Fig. 7.2 Rear wheel assembly (Later quickly detachable type)

1	Quickly-detachable wheel complete	24	Circlip
2	Rim, spokes and hub assembly	25	Bearing sleeve
3	Rear wheel rim (WM2 - 18)	26	Brake anchor plate
4	Spoke (7 9/16 in. 90º) complete with nipple - 20 off	27	Leading brake shoe complete with lining
5	Spoke (7 7/8 in. 90º) complete with nipple - 20 off	28	Trailing brake shoe complete with lining
6	Spoke nipple - 40 off	29	Brake lining - 2 off
7	Hub	30	Brass rivet - 16 off
8	Grease retainer - 2 off	31	Thrust pad - 2 off
9	Backing ring	32	Brake shoe return spring - 2 off
10	Bearing - 2 off	33	Brake cam
11	Dust cap	34	Brake cam lever
12	Bearing sleeve	35	Nut
13	Distance piece (inner)	36	Lever return spring
14	Grease retainer	37	Nut
15	Locking ring	38	Inner distance piece
16	Distance piece (outer)	39	Outer distance piece
17	Plain washer	40	Spindle
18	Nut	41	Rubber sealing ring
19	Sprocket and brake drum (46 teeth)	42	Speedometer drive gearbox
20	Grease retainer	43	Chain adjuster - 2 off
21	Bearing	44	Adjuster end plate - 2 off
22	Felt washer	45	Self-locking nut - 2 off
23	Bearing and felt retainer	46	Tyre security bolt (WM 2)

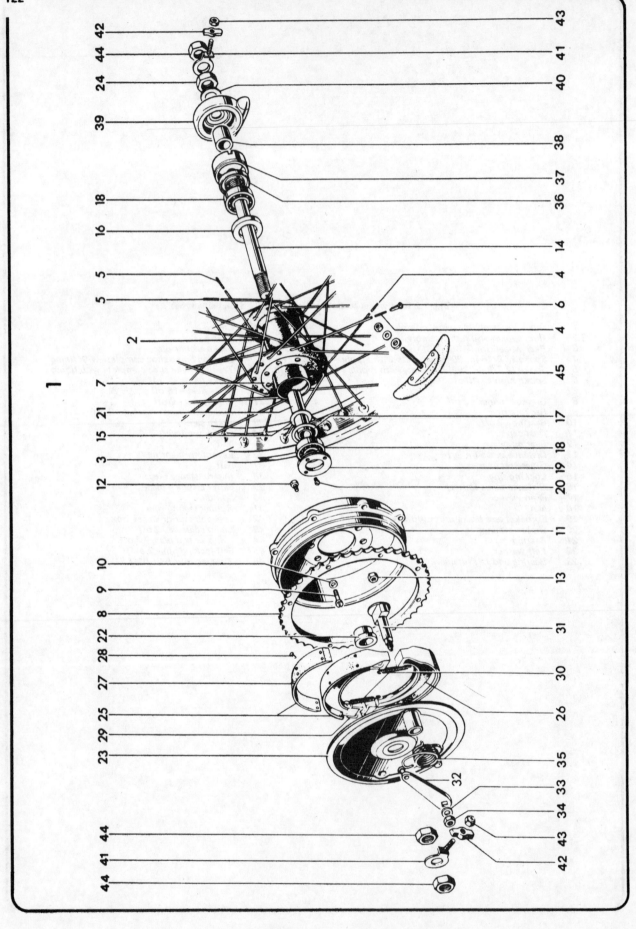

Fig. 7.3 Rear wheel assembly (Later standard type)

1	Rear wheel complete	24	Nut
2	Rim, spokes and hub assembly	25	Leading brake shoe complete with lining
3	Rear wheel rim (WM2 - 18)	26	Trailing brake shoe complete with lining
4	Spoke (7 9/16 in. 90º) complete with nipple - 20 off	27	Brake lining - 2 off
5	Spoke (7 7/8 in. 90º) complete with nipple - 20 off	28	Brass rivet - 16 off
6	Spoke nipple - 40 off	29	Thrust pad - 2 off
7	Hub	30	Brake shoe return spring - 2 off
8	Sprocket - 46 teeth	31	Brake cam
9	Bolt - 8 off	32	Brake cam lever
10	Serrated washer - 8 off	33	Plain washer
11	Brake drum	34	Nut
12	Bolt - 8 off	35	Lever return spring
13	Self-locking nut - 8 off	36	Grease retainer
14	Wheel spindle	37	Speedometer adaptor
15	Grease retainer (brake side)	38	Distance piece
16	Grease retainer	39	Speedometer drive gearbox
17	Backing ring	40	Plain washer
18	Bearing - 2 off	41	Chain adjuster - 2 off
19	Retaining ring	42	Adjuster end plate
20	Locking screw	43	Self-locking nut - 2 off
21	Distance tube	44	Spindle nut - 3 off
22	Distance piece	45	Tyre security bolt (WM 2)
23	Brake anchor plate		

8.2a Drift bearing out of position

8.2b Note use of shim behind bearing

8 Rear brake - removal and examination

1 If the rear wheel is of the standard type or if the rear wheel
has been removed together with the brake drum and sprocket
(quickly detachable wheels) the rear brake assembly is accessible
when the brake plate is lifted away from the brake drum.

2 If the quickly detachable wheel has been removed in the
recommended way (as described in Section 6.6) it will be
necessary to detach the brake drum and sprocket assembly from
the frame. This is accomplished by removing the final drive
chain by detaching the spring link, unscrewing and removing the
nut from the brake plate torque stay so that the latter can be
pulled away, and unscrewing and removing the large nut around
the bearing sleeve that supports the brake and sprocket assembly.
Note that the brake drum has a bearing in its centre which
should be knocked out, cleaned and examined before replacement.

3 The rear brake assembly is similar to that of the front wheel
drum brake apart from the fact that it is of the single leading
shoe type and therefore has only one operating arm. Use an
identical procedure for examining and renovating the brake
assembly to that described in Section 3, paragraphs 3 to 7. Note
that the brake shoes are fitted with thrust pads and not
abutment pads in this instance.

8.3 Rear brake is of single leading shoe type

9 Rear brake - replacement

1 Reverse the dismantling procedure when replacing the front
brake to give the correct sequence of operations. In the case of
the quickly detachable rear wheel, the brake drum and sprocket
should be fitted to the frame first to aid assembly.

2 It is important not to omit the rubber sealing ring of the
quickly detachable wheel which fits over the splines of the hub
and is compressed when the splines are located with those cut
within the brake drum centre. The seal must be in good
condition if it is to prevent the entry of road grit and other
foreign matter which may cause rapid wear of the splines.

10 Rear wheel sprocket - examination

1 When the rear wheel is being worked on it is advisable to
closely inspect the rear sprocket for signs of hooking, erratic
wear or broken teeth.

2 If, on checking, any of these defects are evident, the sprocket
must be replaced. It is preferable to replace the chain at the
same time. Note the rear wheel sprocket is integral with the brake
drum.

3 Examine also the gearbox sprocket. A worn rear sprocket is
usually an indication that the gearbox sprocket is worn too.

11 Final drive chain - examination and lubrication

1 The rear chain should not be forgotten. All too often it is
left until the chain demands attention, by which time it is
usually too late.

2 Every 1000 miles, or 500 miles in winter, rotate the wheel to
the position where the joining link is positioned on the rear wheel
sprocket and slip off the spring link so that the joining link can
be pressed out and the chain separated.

3 It is helpful to have an old chain available which can be
joined to the existing chain and drawn around the gearbox
sprocket until it can be joined with the connecting link. This will
free the existing chain for cleaning. Leave the old chain in place
and use it to feed the cleaned and greased chain back on when
ready.

4 Thoroughly clean the chain with a wire brush and paraffin
(Kerosene). Allow it to drain.

5 To check whether the chain needs renewing, lay it lengthwise
in a straight line and compress it endwise so that all play is taken
up. Anchor one end firmly, then pull endwise in the opposite
direction and measure the amount of stretch. If it exceeds ¼ inch

per foot, renewal is necessary. Never use an old or worn chain when new sprockets are fitted; it is advisable to renew the chain at the same time so that all new parts run together.

6 Every 2000 miles remove the chain and clean it thoroughly in a bath of paraffin before immersing it in a special chain lubricant such as Linklyfe or Chainguard. These latter types of lubricant are applied in the molten state (the chain is immersed) and therefore achieve much better penetration of the chain links and rollers. Furthermore, the lubricant is less likely to be thrown off when the chain is in motion.

7 When replacing the chain, make sure the spring link is positioned correctly, with the closed end facing the direction of travel. Replacement is made easier if the ends of the chain are pressed into the teeth of the rear wheel sprocket whilst the connecting link is inserted, or a simple 'chain-joiner' is used.

12 Final drive chain - automatic lubrication

1 The rear chain is lubricated automatically by an ingenious metering jet which allows small quantities of oil from an oil tank bleed-off to lubricate the rear chain.

2 If the machine is delivering too much or not enough oil to the chain, a certain amount of trial and error may be necessary before the correct adjustment is achieved.

3 Slacken the locknut located in the neck of the oil tank and adjust the bleed-off as necessary.

13 Adjusting the final drive chain

1 Adjusting the rear chain is quite simple. There should be ¾ in of up and down movement at the tightest spot, with the rider seated on the machine.

2 Slacken the wheel nuts and turn the chain adjusting nuts until the right amount of play is achieved. Apply the rear brake and fasten both wheel nuts tightly (re-check).

14 Front brake - adjustment

1 Twin leading shoe brakes are adjusted by a rotating link bar, adjustment of which is usually called for only when new linings have been fitted.

2 Slacken the link rod locking nut, and with the aid of another person, put a spanner on each arm nut and press on both spanners.

3 With the spanners holding each shoe hard against the drum, adjust the length of the arm to approximately 6½ inch centres or when slack in the rod is felt with the clevis pins located.

4 Normal adjustment is quite straightforward. Turn the knurled locknut back at the handlebar lever half a turn, then unscrew the centre adjuster nut until about 1/8 inch slack at the handlebar lever is achieved. Tighten the locknut to stop the adjuster from vibrating loose.

15 Rear brake - adjustment

1 Rear brake adjustment is accomplished by turning the adjuster nut on the end of the brake rod, passing from the brake pedal to the rear brake operating arm.

2 Brake adjustment is also necessary when the rear chain adjustment has been altered.

3 When the rear brake has been adjusted check the stop lamp operation and adjust as necessary.

16 Front wheel balancing

1 It is customary to balance the wheels on high performance machines; an out of balance wheel will be noticed at high speeds by a vibrating wobble transmitted through the handlebars.

2 Set the machine up with the wheels off the ground and spin

10.1 Comparison of old and new sprockets

11.3 Remove spring link to separate chain

11.8 Closed end of spring must always face direction of travel

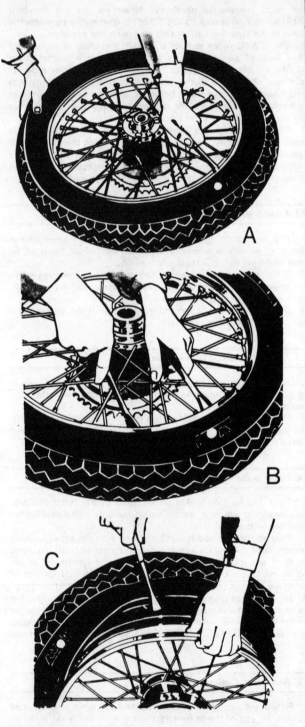

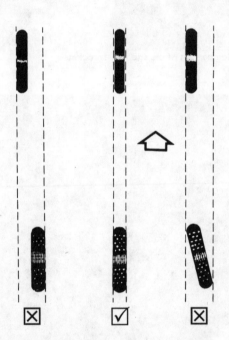

Fig. 7.4 Checking wheel alignment

Fig. 7.5a Tyre removal

A Deflate inner tube and insert lever in close proximity to tyre valve
B Use two levers to work bead over the edge of rim
C When first bead is clear, remove tyre as shown

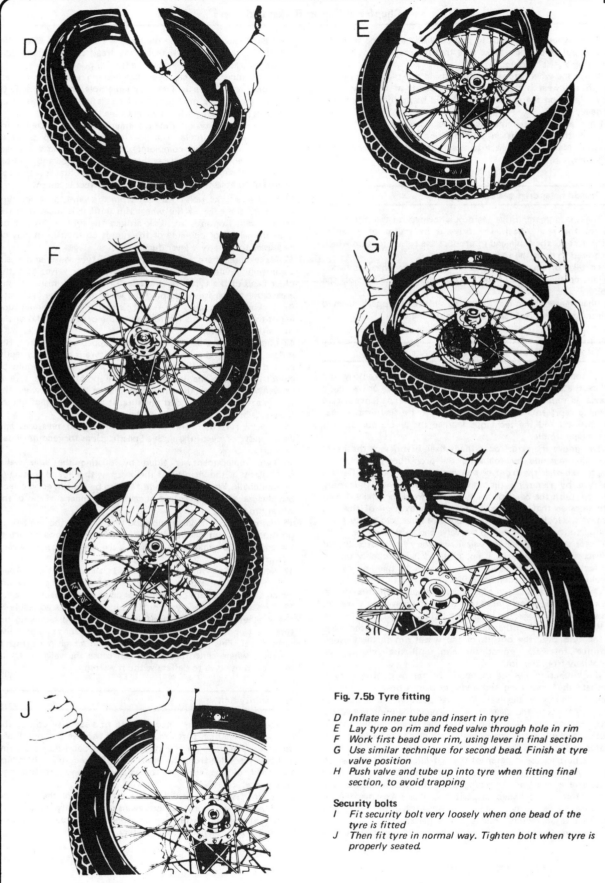

Fig. 7.5b Tyre fitting

D Inflate inner tube and insert in tyre
E Lay tyre on rim and feed valve through hole in rim
F Work first bead over rim, using lever in final section
G Use similar technique for second bead. Finish at tyre valve position
H Push valve and tube up into tyre when fitting final section, to avoid trapping

Security bolts
I Fit security bolt very loosely when one bead of the tyre is fitted
J Then fit tyre in normal way. Tighten bolt when tyre is properly seated.

the front wheel. If it stops in one position each time then the wheel is out of balance. Make sure that the brake is not binding and giving a false reading.

3 Where the wheel stops continuously in one place, the part of the wheel lowest is the heavy part - weights of 1 oz and ½ oz which clamp on the spoke nipple should be positioned opposite the heavy part.

4 Keep adding and removing weights until the wheel can be spun so that it stops in any position. The wheel is now balanced. Make a final check to ensure that the balance weights are secure.

17 Speedometer drive gearbox - general

1 The speedometer drive gearbox is located on the right hand side of the rear wheel. The drive is by means of a square which accepts the cable end to produce the rotary motion which is required to drive the speedometer head.

2 To take off the drive box take out the rear wheel, unfasten the nut on the outside of the gearbox and pull the box off the hub.

3 When reassembling make sure the slots have properly located with the gearbox drive tabs, before tightening up.

18 Tyres - removal and replacement

1 At some time or other the need will arise to remove and replace the tyres, either as the result of a puncture or because a renewal is required to offset wear. To the inexperienced, tyre changing represents a formidable task yet if a few simple rules are observed and the technique learned the whole operation is surprisingly simple.

2 To remove the tyre from either wheel, first detach the wheel from the machine by following the procedure given in this Chapter whether the front or the rear wheel is involved. Deflate the tyre by removing the valve insert and when it is fully deflated, push the bead of the tyre away from the wheel rim on both sides so that the bead enters the centre well of the rim. Remove the locking cap and push the tyre valve into the tyre.

3 Insert a tyre lever close to the valve and lever the edge of the tyre over the outside of the wheel rim. Very little force should be necessary; if resistance is encountered it is probably due to the fact that the tyre beads have not entered the well of the wheel rim all the way round the tyre.

4 Once the tyre has been edged over the wheel rim, it is easy to work around the wheel rim so that the tyre is completely free on one side. At this stage, the inner tube can be removed.

5 Working from the other side of the wheel, ease the other edge of the tyre over the outside of the wheel rim furthest away. Continue to work around the rim until the tyre is free completely from the rim.

6 If a puncture has necessitated the removal of the tyre, re-inflate the inner tube and immerse it in a bowl of water to trace the source of the leak. Mark its position and deflate the tube. Dry the tube and clean the area around the puncture with a petrol soaked rag. When the surface has dried, apply rubber solution and allow this to dry before removing the backing from a patch and applying the patch to the surface.

7 It is best to use a patch of the self-vulcanising type, which will form a very permanent repair. Note that it may be necessary to remove a protective covering from the top surface of the patch, after it has sealed in position. Inner tubes made from synthetic rubber may require a special type of patch and adhesive if a satisfactory bond is to be achieved.

8 Before replacing the tyre, check the inside to make sure that

the agent which caused the puncture is not trapped. Check the outside of the tyre, particularly the tread area, to make sure nothing is trapped that may cause a further puncture.

9 If the inner tube has been patched on a number of past occasions, or if there is a tear or large hole, it is preferable to discard it and fit a new tube. Sudden deflation may cause an accident, particularly if it occurs with the front wheel.

10 To replace the tyre, inflate the inner tube just sufficiently for it to assume a circular shape. Then push it into the tyre so that it is enclosed completely. Lay the tyre on the wheel at an angle and insert the valve through the rim tape and the hole in the wheel rim. Attach the locking cap on the first few threads, sufficient to hold the valve captive in its correct location.

11 Starting at the point furthest from the valve, push the tyre bead over the edge of the wheel rim until it is located in the central well. Continue to work around the tyre in this fashion until the whole of one side of the tyre is on the rim. It may be necessary to use a tyre lever during the final stages.

12 Make sure there is no pull on the tyre valve and again commencing with the area furthest from the valve, ease the other bead of the tyre over the edge of the rim. Finish with the area close to the valve, pushing the valve up into the tyre until the locking cap touches the rim. This will ensure the inner tube is not trapped when the last section of the bead is edged over the rim with a tyre lever.

13 Check that the inner tube is not trapped at any point. Re-inflate the inner tube, and check that the tyre is seating correctly around the wall of the tyre on both sides, which should be equidistant from the wheel rim at all points. If the tyre is unevenly located on the rim, try bouncing the wheel when the tyre is at the recommended pressure. It is probable that one of the beads has not pulled clear of the centre well.

14 Always run the tyres at the recommended pressures and never under or over-inflate. See Specifications for recommended pressures.

15 Tyre replacement is aided by dusting the side walls, particularly in the vicinity of the beads, with a liberal coating of French chalk. Washing up liquid can also be used to good effect, but this has the disadvantage of causing the inner surfaces of the wheel rim to rust.

16 Never replace the inner tube and tyre without the rim tape in position. If this precaution is overlooked there is good chance of the ends of the spoke nipples chafing the inner tube and causing a crop of punctures.

17 Never fit a tyre which has a damaged tread or side walls. Apart from the legal aspects, there is a very great risk of a blow-out, which can have serious consequences on any two wheel vehicle.

18 Tyre valves rarely give trouble but it is always advisable to check whether the valve itself is leaking before removing the tyre. Do not forget to fit the dust cap which forms an effective second seal. This is especially important on a high performance machine, where centrifugal force can cause the valve insert to retract and the tyre to deflate without warning.

19 Security bolt

1 It is often considered necessary to fit a security bolt to the rear wheel of a high performance model because the initial take up of drive may cause the tyre to creep around the wheel rim and tear the valve from the inner tube. The security bolt retains the bead of the tyre to the wheel rim and prevents this occurrence.

2 A security bolt is fitted to the rear wheel of the Triumph 350/500 cc unit-construction twins as a safety precaution. Before attempting to remove or replace the tyre, the security bolt must be slackened off completely.

20 Fault diagnosis

Symptom	Reason/s	Remedy
Handlebars oscillate at low speeds	Buckle or flat in wheel rim, most probably front wheel	Check rim alignment by spinning wheel. Correct by retensioning spokes or rebuilding on new rim.
	Tyre not straight on rim	Check tyre alignment.
Machine lacks power and accelerates poorly	Brakes binding	Warm brake drum provides best evidence. Re-adjust brakes.
Brakes grab when applied gently	Ends of brake shoes not chamfered	Chamfer with file.
	Elliptical brake drum	Lightly skim in lathe (specialist attention required).
Front brake feels spongy	Air in hydraulic system (disc brake only)	Bleed brake.
Brake pull-off sluggish	Brake cam binding in housing	Free and grease.
	Weak brake shoe springs	Renew if springs have not become displaced.
	Sticking pistons in brake caliper (front disc brake only)	Overhaul caliper unit.
Harsh transmission	Worn or badly adjusted final drive chain	Adjust or renew as necessary.
	Hooked or badly worn sprockets	Renew as a pair.
	Loose rear sprocket (standard wheel only)	Check sprocket retaining bolts.

Chapter 8 Electrical system

Contents

Specifications

Model T100R — Daytona

12 volt electrical system

Battery	1 Lucas 12 volt battery PUZ5A or earlier 2 Lucas 6 volt batteries connected in series (MKZ9E)
Rectifier type	Lucas 2DS506
Alternator type	Lucas RM19
Horn	Clear hooter 27899 12 volt

Bulbs:	No	Type
Headlight	370	45/35W (vert dip)
Parking light	989	5W MCC
Stop and tail light	380	21/5W offset pin
Speedometer	987	3W MES
Oil pressure warning light	281	2W (WL15)
Hi-beam warning light	281	2W (WL15)
Direction indicator warning light	281	2W (WL15)
Flashing indicators	382	

Zener diode type	ZD 715
Coil type	Lucas 17M12 (12v) 2 off
Contact breaker type	Lucas 6CA (12° range)
Fuse rating	35 amp

Model T100S — Tiger 100

Bulbs:	No	Type
Headlight	Lucas 414	50/40 watts pre-focus
Parking light	Lucas 989	6 watts MCC
Stop and tail light	Lucas 380	6/21 watts offset pins
Speedometer light	Lucas 987	2 watt MES
Main beam indicator (where fitted)	Lucas 281	2 watt (BA7S)
Ignition warning light	Lucas 281	2 watt (BA7S)

Coil type	Lucas 17M12 (12 v) 2 off
Contact breaker type	Lucas 4CA (12° range)
	After H 57083 Lucas 6CA (12° range)

Model T90 — Tiger 90

Battery type	1 Lucas 12 volt battery (PUZ5A) or alternatively 2 Lucas 6 volt batteries connected in series (MKZ9E)
Rectifier type	Lucas 2DS506
Alternator type	Lucas RM19
Horn	Clear hooter 27899 (12 volt)

Bulbs:	No	Type
Headlight	Lucas 414	50/40 watts pre-focus
Parking light	Lucas 989	6 watts MCC
Stop and tail light	Lucas 380	6/21 watts offset pins
Speedometer light	Lucas 987	2 watts MES
Main beam indicator light (where fitted)	Lucas 281	2 watt (BA7S)
Ignition warning light	Lucas 281	2 watt (BA7S)

Coil type	Lucas MA12 (2 off)
Contact breaker type	Lucas 6CA
Fuse rating	35 amp

Model 5TA — Speed Twin

Battery type	Lucas 12 volt (type PUZ5A) or alternatively 2 Lucas 6 volt (type MKZ9E) connected in series
Rectifier type	Lucas 2DS506
Alternator type	Lucas RM19
Horn	Lucas 8H (12 volt)

Bulbs (6 volt):	No	Type
Headlight	Lucas 414	50/40 watts pre-focus
Parking light	Lucas 989	6 watts MCC
Stop and tail light	Lucas 380	6/21 watts offset pins
Speedometer light	Lucas 987	2 watts MES

Coil type	Lucas MA12 (12 volt)
Contact breaker type	Lucas 4CA
Fuse rating	35 amp

Model 3TA — Twenty One

Battery type	1 Lucas 12 volt battery (PUZ5A) or alternatively 2 Lucas 6 volt batteries connected in series (MKZ9E)
Rectifier type	2DS 506
Alternator type	RM19
Horn	Lucas 8H (12 volt)

Bulbs:	No	Type
Headlight	Lucas 414	50/40 watt pre-focus
Parking light	Lucas 989	6 watts MCC
Stop and tail light	Lucas 380	6/21 watts offset pins
Speedometer light	Lucas 987	2 watts MES

Coil type	Lucas MA12 (2 off)
Contact breaker type	Lucas 4CA
Fuse rating	35 amp

Electrical equipment AC magneto (ET) ignition equipment

Alternator type	RM19 ET
Horn type	Clear hooter A585 (6 volt)
Coil type	3 ET
Condensers (capacitors)	Lucas 54441582
Contact breaker type	Lucas 4CA

Bulbs (6 volt):	No	Type
Headlight	Lucas 312	30/24 watts pre-focus
Stop and tail light	Lucas 384	6/18 watts offset pins

Unless otherwise stated, T100R data applies to all models

1 General description

The electrical system is supplied by an alternator situated on the end of the crankshaft. The output is converted into the direct current by a silicon diode rectifier and then supplied to a 12 volt battery.

An electrical device known as a Zener diode regulates the charge rate to suit the condition of the battery. The ignition system, as described in Chapter 5, derives its supply from the rectified current.

There is no emergency start as the charge from the alternator is sufficient to overcome a fully discharged battery and produce a spark.

On machines with an engine number up to H40528, there is an emergency start facility. By switching the ignition key to the emergency start position marked on the switch the machine should then start. A change over to the normal switch position should be made as soon as possible or damage to the points will occur.

2 Battery - charging procedure and maintenance

1 Whilst the machine is used on the road it is unlikely that the battery will require attention other than routine maintenance because the generator will keep it fully charged. However, if the machine is used for a succession of short journeys only, mainly in darkness when the lights are permanently in use, it is possible that the output from the generator may fail to keep pace with the heavy electrical demand, especially if the machine is parked with the lights switched on. In such cases it will be necessary to remove the battery from time to time to have it charged independently.

2 The battery is located below the dualseat, in a carrier slung between the two parallel frame tubes. It is secured by a strap which, when released, will permit the battery to be withdrawn after disconnection of the leads. The battery positive is always earthed. To remove the battery carrier, release the earth lead and rectifier by unscrewing the retaining nut. Slacken the nuts on the cross-straps and lift the carrier away.

3 The normal charge rate is 1 amp. A more rapid charge can be given in an emergency, but this should be avoided if possible because it will shorten the battery's life.

4 When the battery is removed from the machine, detach the cover and clean the battery top. If the terminals are corroded, scrape them clean and cover them with Vaseline (not grease) to protect them from further attack. If a vent tube is fitted, make sure it is not obstructed and that it is arranged so that it will not discharge over any parts of the machine.

5 If the machine is laid up for any period of time, the battery should be removed and given a 'refresher' charge every six weeks or so in order to maintain it in good condition.

6 When two 6 volt batteries are fitted, they must be connected in series with one another. The negative of one battery must go to the wiring harness and the positive of the OTHER to the frame or earth connection. The intermediate connection is made by joining the free negative terminal of one battery to the free positive terminal of the other.

3 Silicon diode rectifier - general

1 The silicon diode rectifier is bolted to a bracket attached to the rear of the battery carrier, beneath the dualseat. Its function is to convert the alternating current from the alternator to direct current, which can then be used to charge the battery and operate the ignition circuit.

2 The rectifier is deliberately placed in this location so that it is not exposed directly to water or oil and yet has free circulation of air to permit cooling. It should be kept clean and dry; the nuts connecting the rectifier plates should not be disturbed under any circumstances.

3 It is not possible to check whether the rectifier is working properly without the necessary test equipment. If performance is suspect, a Triumph agent or auto-electrical expert should be consulted. Note that the rectifier will be destroyed if it is subjected to a reverse flow of current.

4 When tightening the rectifier securing nut, hold the nut at the other end with a spanner. Apart from the fact that the securing stud is sheared very easily if overtightened, there is a risk of the plates twisting and severing their internal connections.

4 Zener diode - general

1 Only the 12 volt systems have Zener diodes. This is located under the headlamp in a heat sink, so called because excess voltage is returned into heat and then the heat sink absorbs heat and dissipates it through the fins.

2 If the diode is suspected of failure a quick check should be made to see if it will pass current both ways. If it does, the diode has broken down and must be renewed. If in doubt take it to a qualified electrical expert to check.

3 When refitting the Zener diode, take great care not to over-tighten the mounting as the diode case is made of copper and is extremely soft. Tighten to a torque setting of 1½ lb ft (0.207 kg m).

2.1 Battery is located under dualseat, across frame

4.1 Zener diode is mounted on lower fork yoke

5 Headlamp - replacing bulbs and adjusting beam height

1   To replace the bulbs slacken the screw in the top of the headlight and, using both hands to support the unit, pull away at the top. When the top comes away from the shell, unhook the bottom tab.

2   Hold the headlamp reflector unit in one hand and with the other depress the cap immediately behind the main bulb whilst turning it in an anticlockwise direction.

3   The bulb can now be lifted out. When replacing the main headlamp bulb a location in the form of a half-round cutaway in the bulb flange will line up with the half-round ridge in the bulb sleeve.

4   The bulb cap can be replaced in only one position as the three location pegs are offset.

5   The pilot bulb holder is a push-fit into the base of the reflector. The bulb has a bayonet fitting.

6   Re-locate the tab at the bottom of the headlamp rim and press the reflector unit back in place. Tighten the retaining screw.

7   To adjust beam height, sit on the machine with the centre stand retracted and turn on the headlamp. Flick on to the main beam and taking the headlamp as the correct height, push the whole headlamp assembly forward until the main beam is concentrated in an area slightly lower than the height of the headlamp.

8   If indicators are fitted, slacken the stems slightly and adjust as before. Retighten but do not overtighten as the stems are hollow and will shear relatively easily.

6 Tail and stop lamp - replacing bulb

1   The combined tail and stop lamp is fitted with a double filament bulb with offset pins to prevent its unintentional reversal in the bulb holder. The lamp unit serves a two-fold purpose; to illuminate the rear of the machine and the rear number plate, and to give visual warning when the rear brake is applied. To gain access to the bulb, remove the two screws securing the plastic lens cover in position. This has a sealing gasket below it. The bulb is released by pressing inward with a twisting action; it is rated at 5/21 W, 12 volts.

2   The stop lamp is actuated by a switch bolted to the rear chainguard. The switch is connected to the brake rod by a spring attached to a clamp fitting around the rod; the point at which the switch operates is governed by the position of the clamp in relation to the rod. If the clamp is moved forward, the stop lamp will indicate earlier and vice versa. The switch does not require attention other than the occasional drop of thin oil.

7 Speedometer and tachometer bulbs - replacement

The speedometer and tachometer heads are each fitted with a bulb to illuminate the dial when the headlamp is switched on. The bulb holders are a push fit into the bottom of each instrument case and carry a 3W, 12 volt bulb which has a threaded body.

8 Ignition warning, oil pressure warning, main beam and flashing indicator warning bulbs - replacement

1   The combination of lamps fitted varies according to the model and year of manufacture. There is not necessarily four separate forms of warning fitted to each machine.

2   The bulb holders are a push fit into either the top or the back of the headlamp shell, depending on the type of headlamp fitted. Each of the bulbs is rated at 2W, 12 volts.

5.1 Slacken screw to release headlamp rim and reflector unit

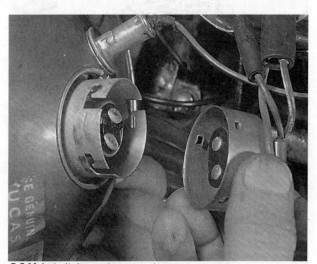
5.2 Main bulb has a plug-on socket

5.3 Locating pip prevents misalignment of main headlamp bulb

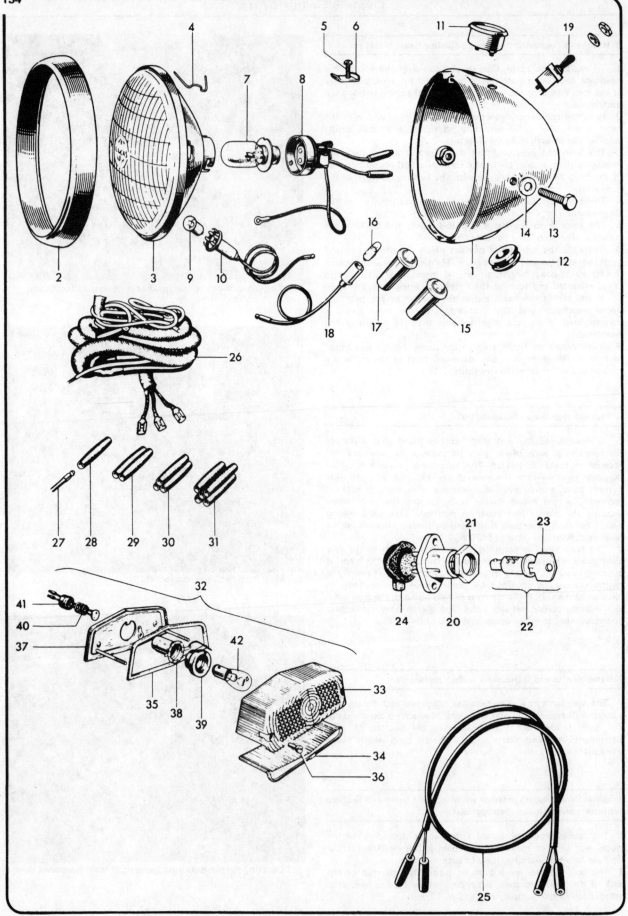

FIG. 8.1. HEADLAMP, WIRING HARNESS, SWITCHES AND TAIL LAMP

1	Headlamp complete, Lucas type SS700P		22	Ignition lock, complete with keys
2	Headlamp rim		23	Ignition key
3	Reflector unit		24	Ignition switch cover
4	Fixing wire clips - 6 off		25	Stop and tail lamp
5	Screw		26	Wiring harness
6	Plate		27	Snap connector terminal
7	Bulb (50/40W, vertical dip Pre-focus)		28	Single snap connector
8	Main bulbholder contacts		29	Double snap connector
9	Pilot bulb (6W. M.C.C.)		30	Triple snap connector
10	Pilot bulbholder		31	Quintuple snap connector
11	Ammeter		32	Stop and tail lamp, Lucas 564
12	Grommet		33	Stop lamp lens
13	Bolt - 2 off		34	Window
14	Plain washer - 2 off		35	Lens sealing gasket
15	Warning light body (red)		36	Sleeve nut - 2 off
16	Bulb (2W Lucas BA7S)		37	Base assembly
17	Warning light body (green)		38	Bulbholder
18	Warning light bulbholder - 2 off		39	Grommet
19	Lighting switch, Lucas 57SA		40	Contact assembly
20	Ignition switch, Lucas S45		41	Grommet
21	Nut		42	Bulb (6/21W, offset pin)

9 Flashing indicator lamps

1 Late models have flashing direction indicator lamps attached to the front and rear of the machine. They are operated by a thumb switch on the left hand end of the handlebars. An indicator lamp built into the rear of the headlamp shell will flash in unison with the lights, provided the front and rear lights are operating correctly.
2 The bulbs are fitted by removing the plastic lens covers, held in position by two screws. The bulbs are of the bayonet type and must be pressed and turned to release or fit. Each bulb is rated at 21W, 12 volts.

10 Flasher unit - location and replacement

1 The flasher unit is located beneath the dualseat, along with the other electrical equipment. It seldom gives trouble unless it is subjected to a heavy blow which will disturb its sensitive action.
2 It is not possible to renovate a malfunctioning flasher unit. If the bulbs are in working order and will give only a single flash when the handlebar switch is operated, the flasher unit should be suspected and, if necessary, renewed.

11 Headlamp dip switch

1 The headlamp dip switch forms part of the switch unit fitted to the right hand side of the handlebars on all late models. Earlier models have a separate dip switch mounted on the left hand side of the handlebars which also contains the horn push.
2 If the dip switch malfunctions, the switch unit must be renewed since it is seldom practicable to effect a satisfactory repair.

12 Horn push and horn - adjustment

1 The horn push on late models forms part of the switch unit at the right hand end of the handlebars. On earlier models, it is combined with the separate dip switch.
2 The horn is secured below the nose of the petrol tank, facing in a forward direction. It is provided with adjustment in the form of a serrated screw inset into the back of the horn body.
3 To adjust the horn, turn the screw anticlockwise until the horn just fails to sound, then back it off about one-quarter turn. Adjustment is needed only very occasionally, to compensate for wear of the internal moving parts.

13 Fuse - location and replacement

1 A fuse is incorporated in the brown/blue coloured lead from the negative terminal of the battery. It is housed within a quickly detachable shell and protects the electrical equipment from accidental damage if a short circuit should occur.
2 If the electrical system will not operate, a blown fuse should be suspected, but before the fuse is renewed, the electrical system should be inspected to trace the reason for the failure of the fuse. If this precaution is not observed, the replacement fuse may blow too.
3 The fuse is rated at 35 amps and at least one spare should always be carried. In an extreme emergency, when the cause of the failure has been rectified and if no spare is available, a get-you-home repair can be effected by wrapping silver paper around the blown fuse and re-inserting it in the fuse holder. It must be stressed that this is only an emergency measure and the 'bastard' fuse should be replaced at the earliest possible opportunity. It affords no protection whatsoever to the electrical circuit when bridged in this fashion.

14 Ignition switch

1 The ignition switch is fitted to the left hand top cover of the forks, or in the left hand or right hand cover surrounding the rear portion of the frame, depending on the model and year of manufacture.
2 It is retained by a locknut or a locking ring which, when unscrewed, will free the switch.

15 Headlamp switch

1 A two or three position headlamp switch is fitted to the headlamp shell to operate the pilot and main headlamp bulbs. Machines fitted with the two-position rotary switch must have the ignition switch in position 4 before the headlamp will operate.
2 Late models have an additional headlamp flasher in the form of a push button embodied in the switch assembly on the right hand end of the handlebars.

16 Ammeter

1 All models with the full headlamp shell (except those fitted with AC ignition) have an ammeter inset into the top of the headlamp shell to show the amount of charge from the generator.
2 When an ammeter is not fitted, as in the case of the late models with the 'short' headlamp shell, an ignition warning light is fitted. This light will not extinguish after the engine is started, if the generator has failed.

17 Capacitor ignition

1 A capacitor is built into the ignition circuit so that a machine without lights (and therefore without a battery) or one on which the battery has failed, can be started and run normally. If lights are fitted, the machine can be used during the hours of darkness, since the lighting equipment will function correctly immediately the engine starts, even if the battery is removed.
2 The capacitor is fitted into a coil spring to protect it from vibration. It is normally mounted with its terminals pointing downward from a convenient point underneath the dualseat.
3 Before running a machine on the capacitor system with the battery disconnected, it is necessary to tape up the battery negative so that it cannot reconnect accidentally and short circuit. If this occurs, the capacitor will be ruined. A convenient means of isolating the battery is to remove the fuse.

18 Wiring - layout and examination

1 The cables of the wiring harness are colour-coded and will correspond with the accompanying wiring diagrams.
2 Visual inspection will show whether any breaks or frayed outer coverings are giving rise to short circuits which will cause the main fuse to blow. Another source of trouble is the snap connectors and spade terminals, which may make a poor connection if they are not pushed home fully.
3 Intermittent short circuits can sometimes be traced to a chafed wire passing through, or close to, a metal component, such as a frame member. Avoid tight bends in the cables or situations where the cables can be trapped or stretched, especially in the vicinity of the handlebars or steering head.

19 Front brake stop lamp switch

1 In order to comply with traffic requirements in certain overseas countries, a stop lamp switch is now incorporated in the front brake cable so that the rear stop lamp is illuminated

when the front brake is applied. When an hydraulic disc brake is fitted, the stop lamp switch is incorporated in the master cylinder unit, clamped to the right hand end of the handlebars.

2 There is no means of adjustment for the front brake stop lamp switch. If the switch malfunctions, the front brake cable must be replaced, or in the case of the disc brake assembly, the switch unit unscrewed from the master cylinder and renewed.

20 Fault diagnosis

Symptom	Cause	Remedy
Complete electrical failure	Blown fuse	Check wiring and electrical components for short circuit before fitting new 15 amp fuse.
	Isolated battery	Check battery connections, also whether connections show signs of corrosion.
Dim lights and horn inoperative	Discharged battery	Recharge battery with battery charger. Check whether generator is giving correct output.
Constantly blowing bulbs	Vibration, poor earth connection	Check security of bulb holders. Check earth return connections.

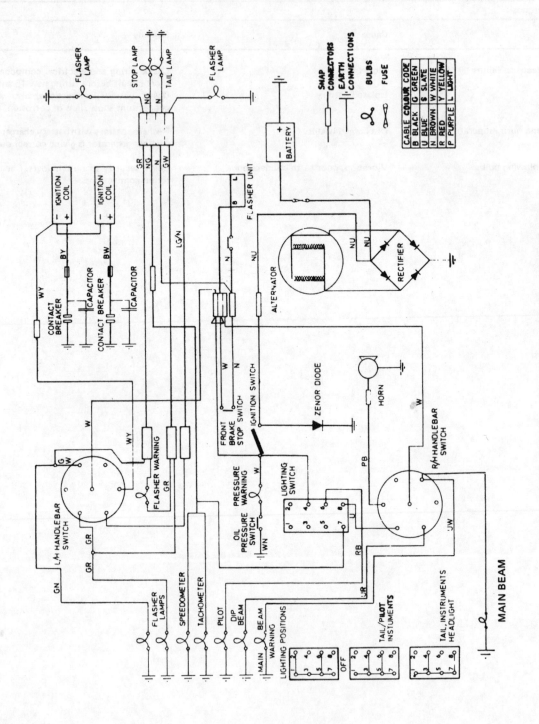

Fig. 8.2 Wiring diagram (UK) 12 volt coil ignition KE 00001 onwards

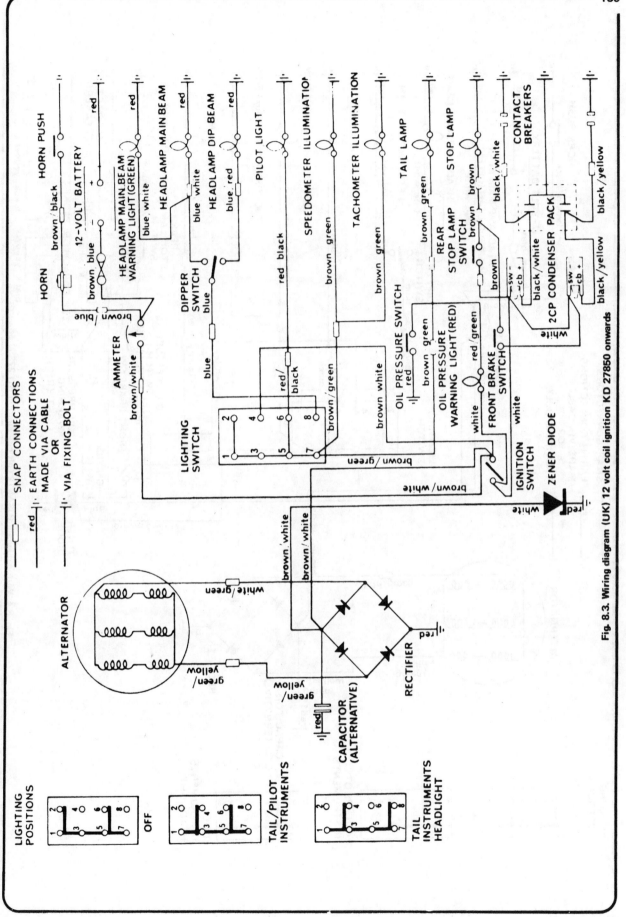

Fig. 8.3. Wiring diagram (UK) 12 volt coil ignition KD 27850 onwards

140

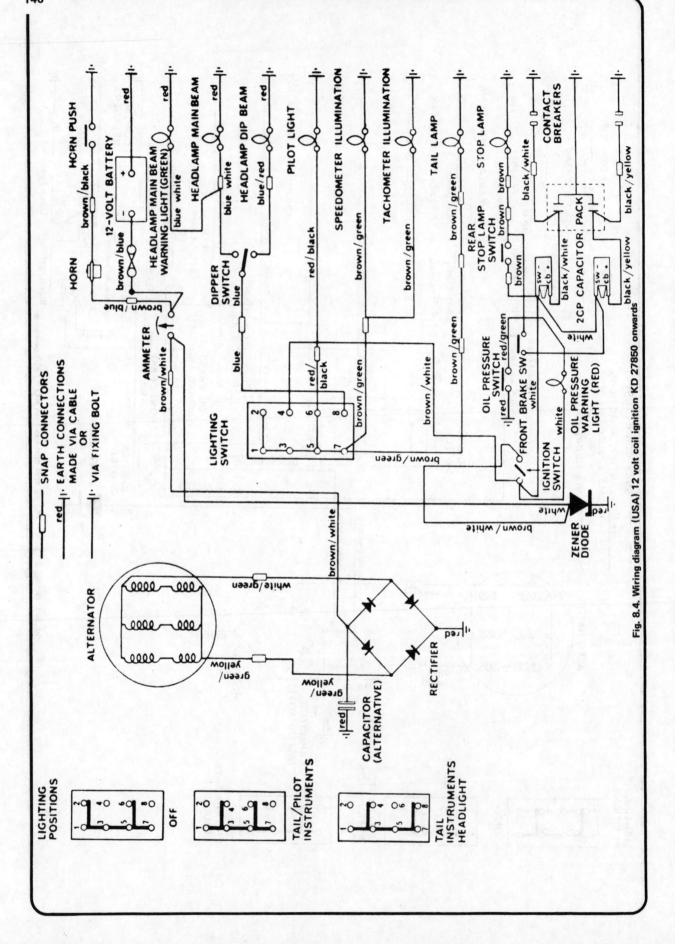

Fig. 8.4. Wiring diagram (USA) 12 volt coil ignition KD 27850 onwards

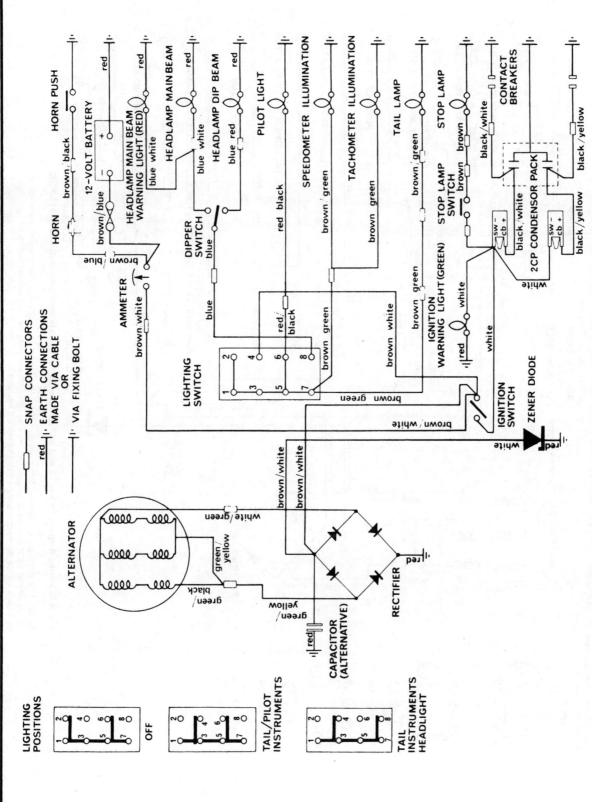

Fig. 8.5. Wiring diagram (UK) 12 volt coil ignition, with separate headlamp H 57083 onwards

142

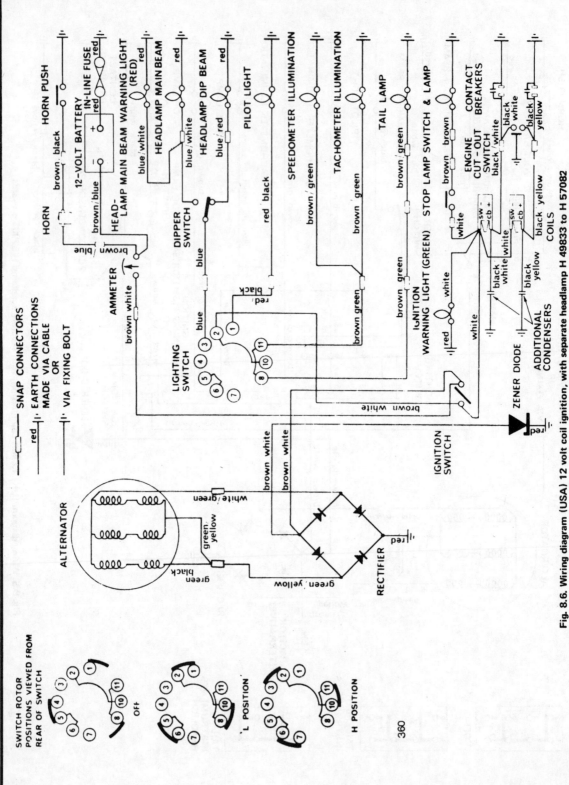

Fig. 8.6. Wiring diagram (USA) 12 volt coil ignition, with separate headlamp H 49833 to H 57082

Note: The main beam warning light (when fitted) is connected to the headlamp main beam wire (blue/white) by a double snap connector. The ignition warning lamp is connected to an ignition coil by a white wire incorporated in the wiring harness

360

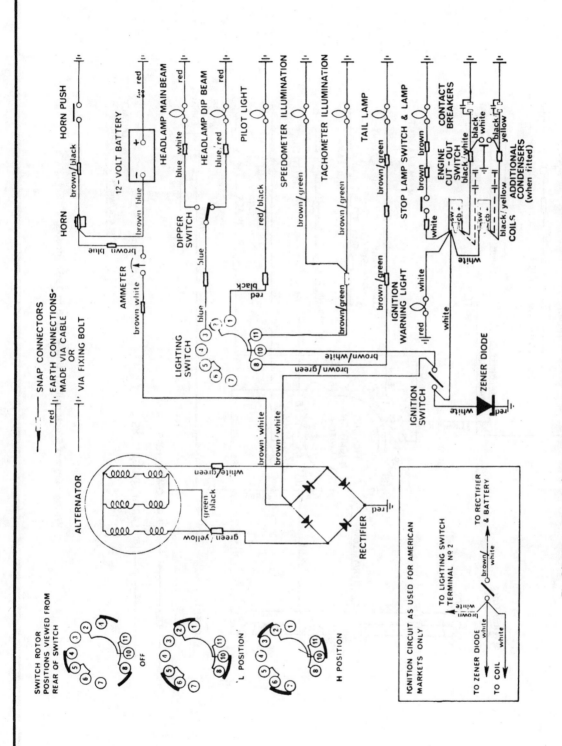

Fig. 8.7. Wiring diagram (UK and Export) 12 volt coil ignition, without nacelle. Up to H 49832

Note: The main beam warning lamp (when fitted) is connected to the headlamp main beam wire (blue/white) by a double snap connector. The ignition warning lamp is connected to an ignition coil by a white wire incorporated in the wiring harness

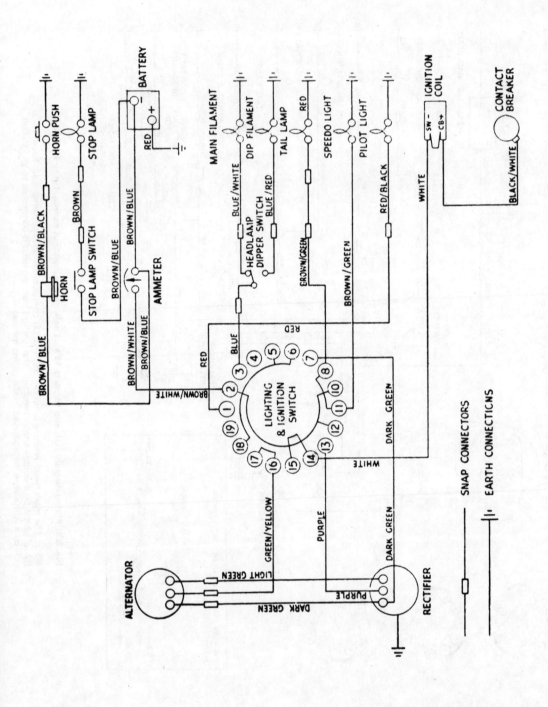

Wiring diagram 6 volt coil ignition – 1958 to 1963 models

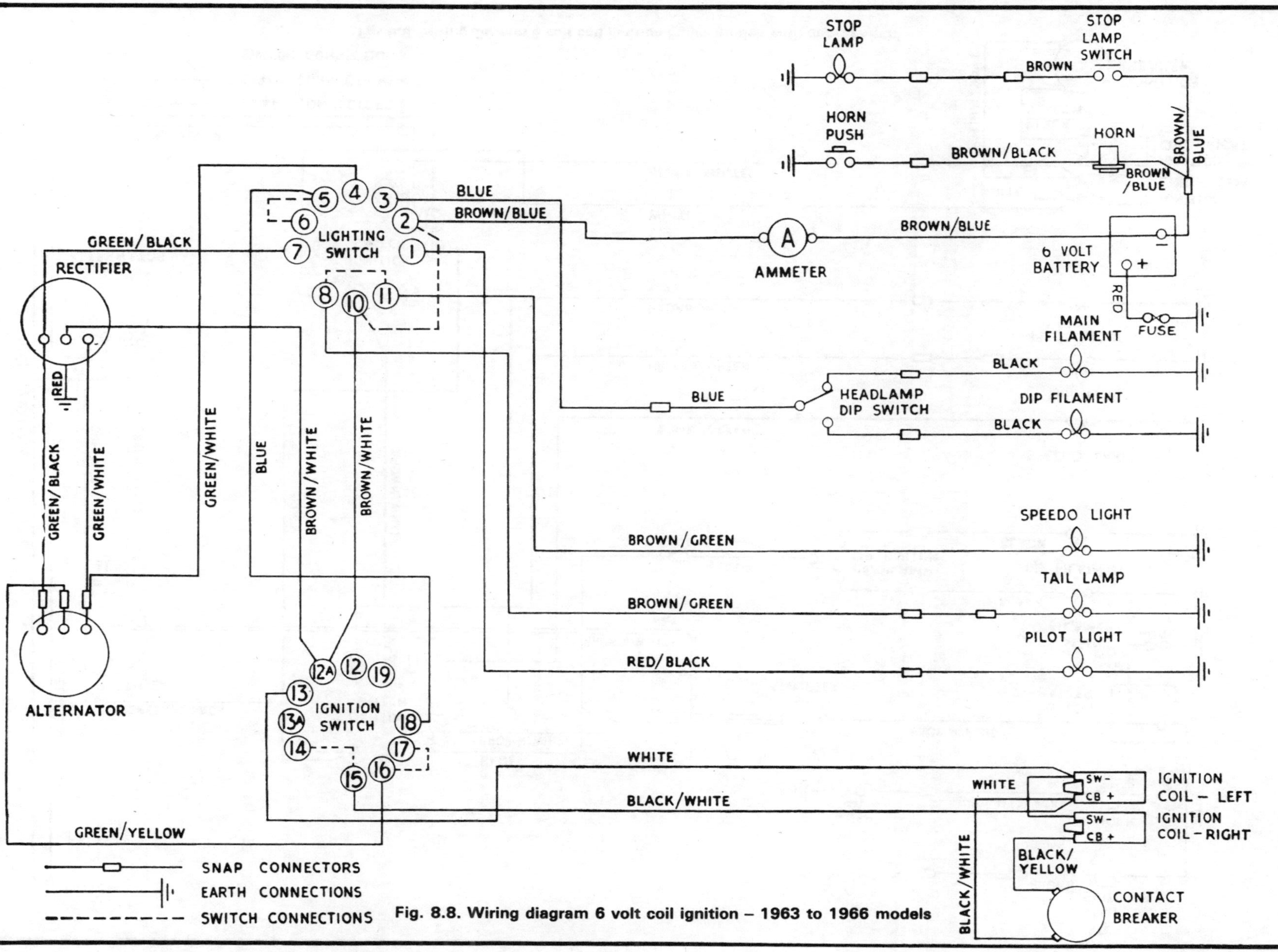

Fig. 8.8. Wiring diagram 6 volt coil ignition – 1963 to 1966 models

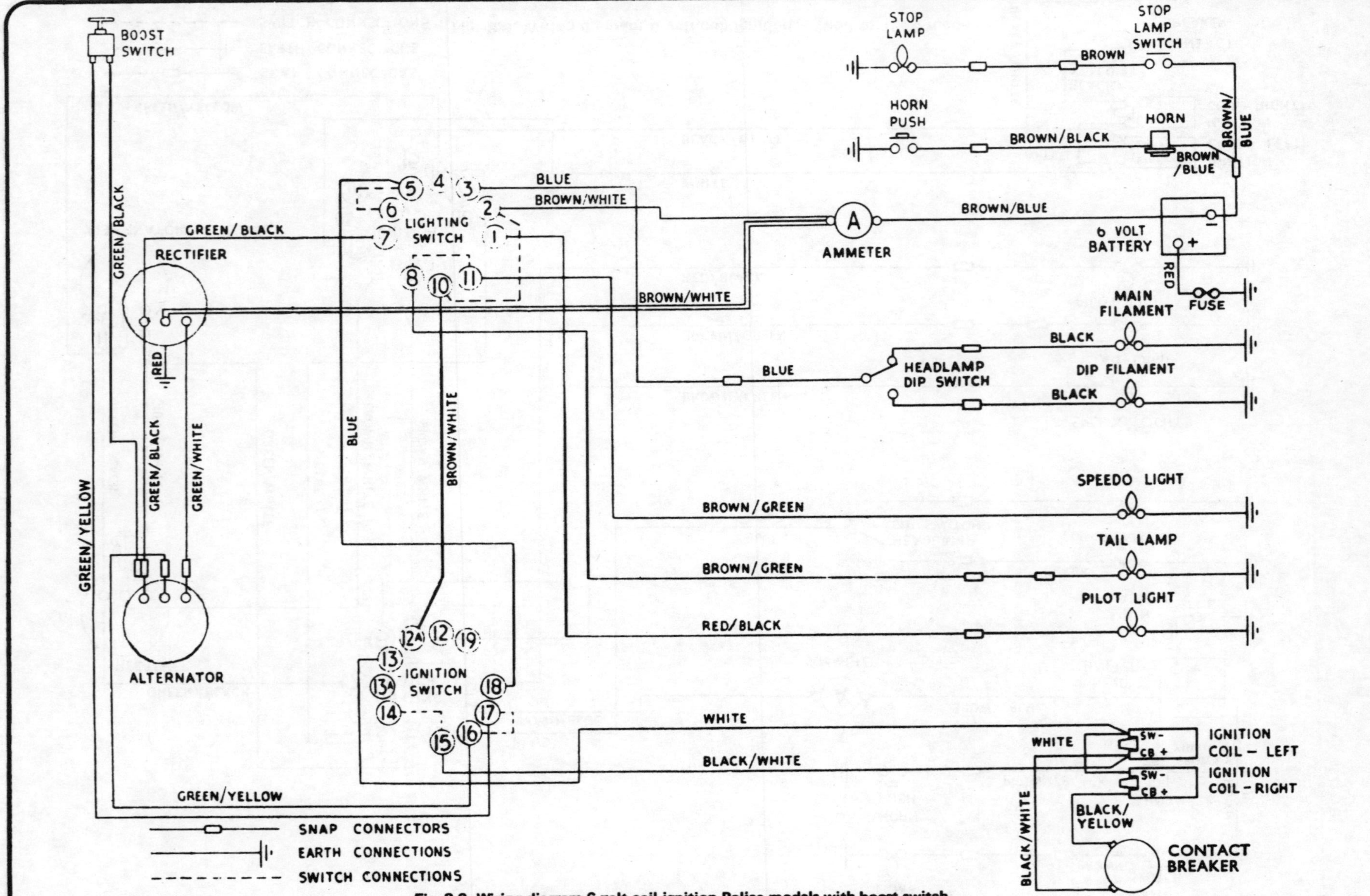

Fig. 8.9. Wiring diagram 6 volt coil ignition Police models with boost switch

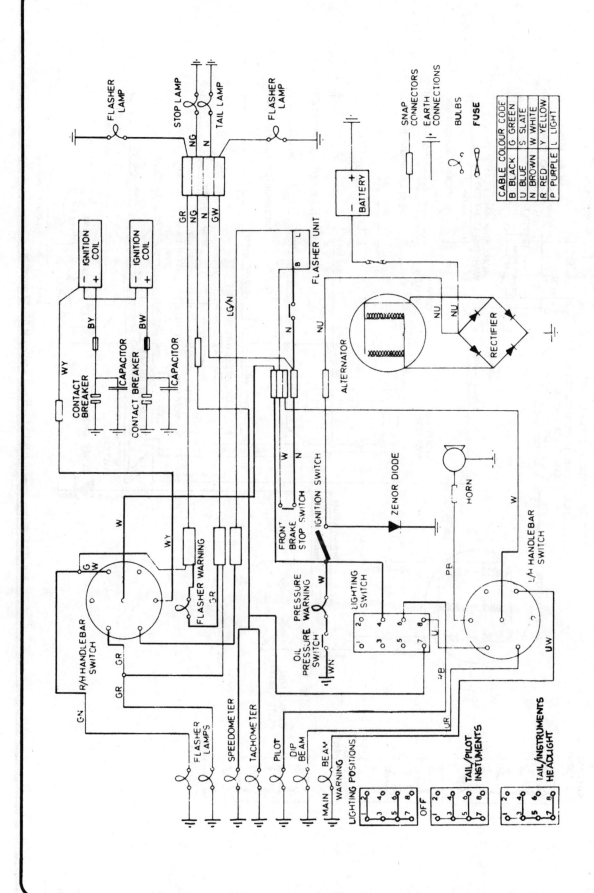

Fig. 8.10. Wiring diagram (USA) 12 volt coil ignition KE 00001 onwards

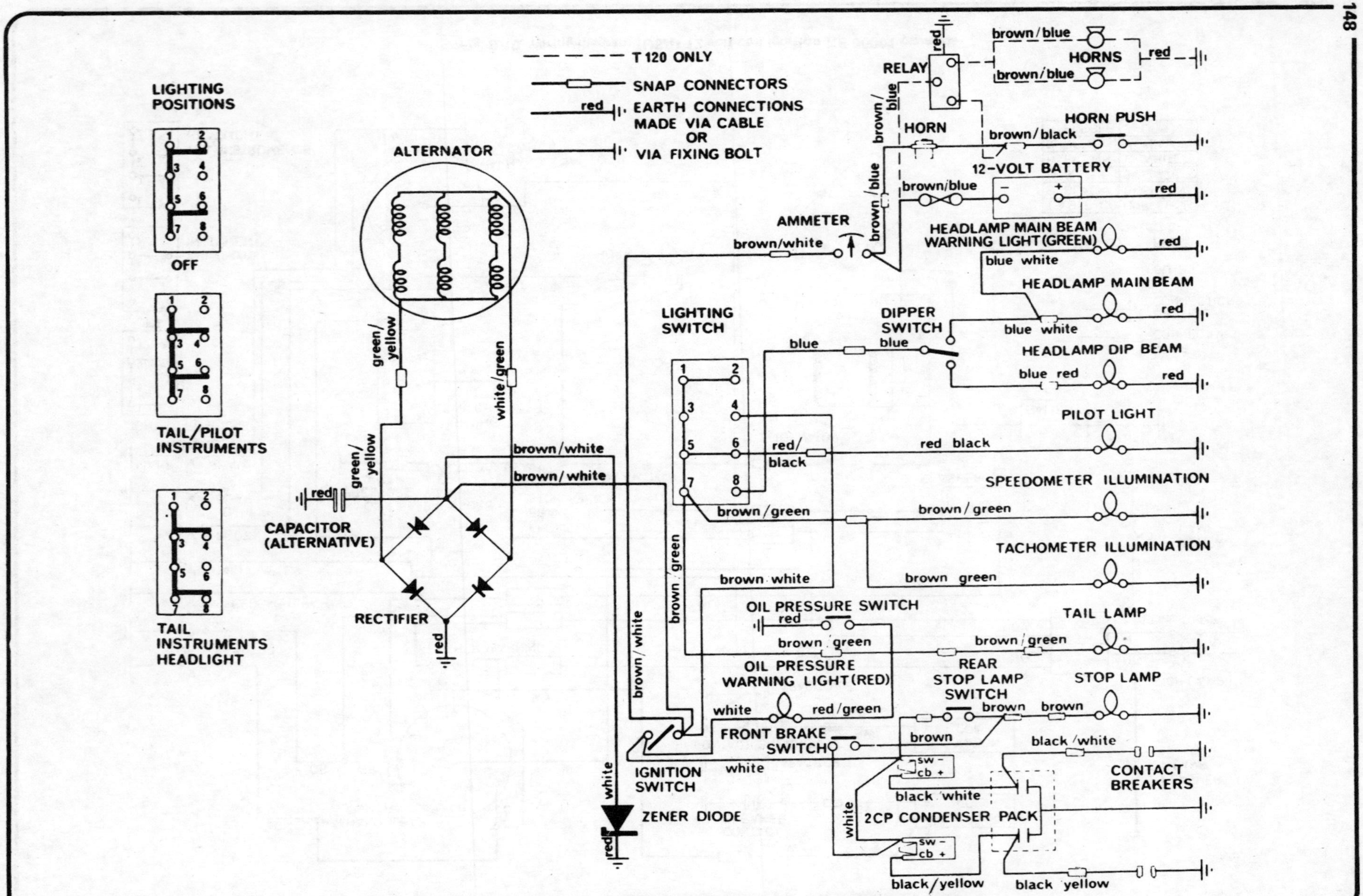

Fig. 8.11. Wiring diagram (UK) 12 volt coil ignition H 65573 onwards

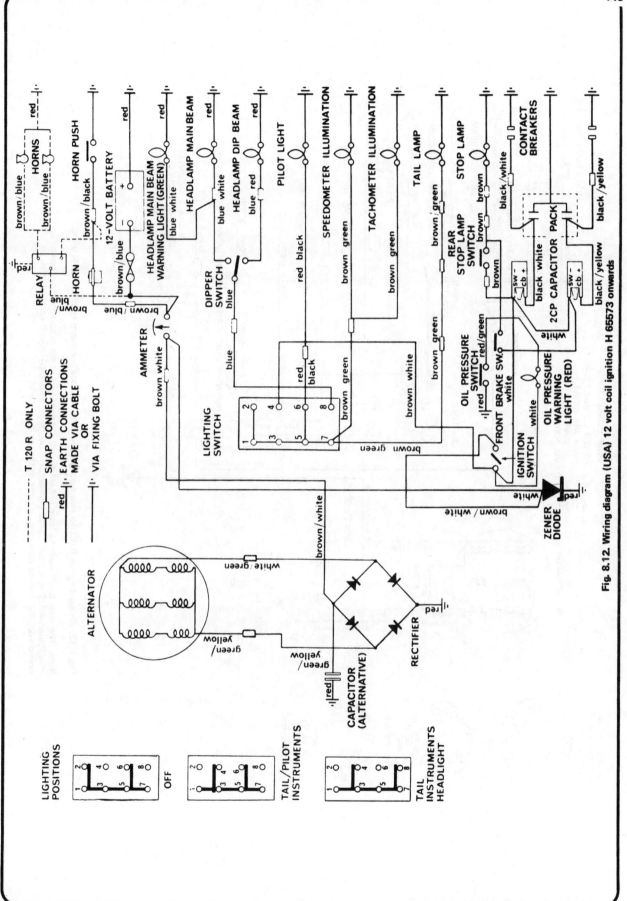

Fig. 8.12 Wiring diagram (USA) 12 volt coil ignition H 65573 onwards

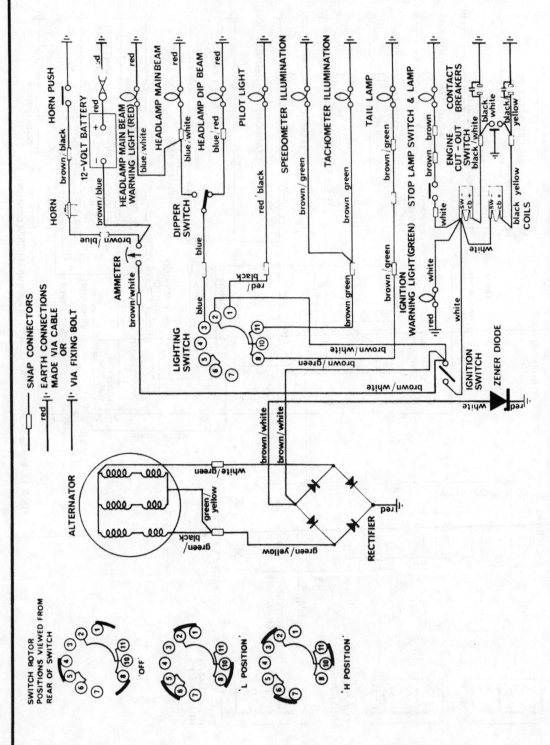

Fig. 8.13. Wiring diagram (UK) 12 volt coil ignition with separate headlamp H 49833 to H 57082

Note: The main beam warning lamp (where fitted) is connected to the headlamp main beam wire (blue/white) by a double snap connector. The ignition warning lamp is connected to an ignition coil by a white wire incorporated in the wiring harness

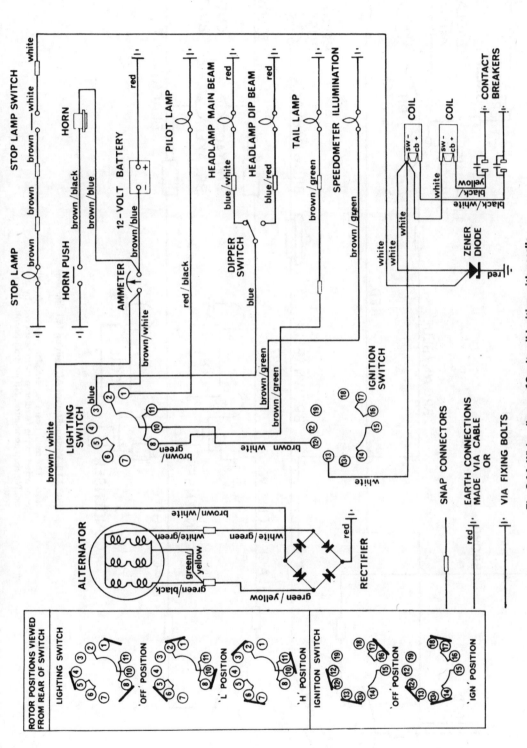

Fig. 8.14. Wiring diagram 12 volt coil ignition with nacelle

Note: The main beam warning lamp (where fitted) is connected to the headlamp main beam wire (blue/white) by a double snap connector. The ignition warning lamp is connected to an ignition coil by a white wire incorporated in the wiring harness

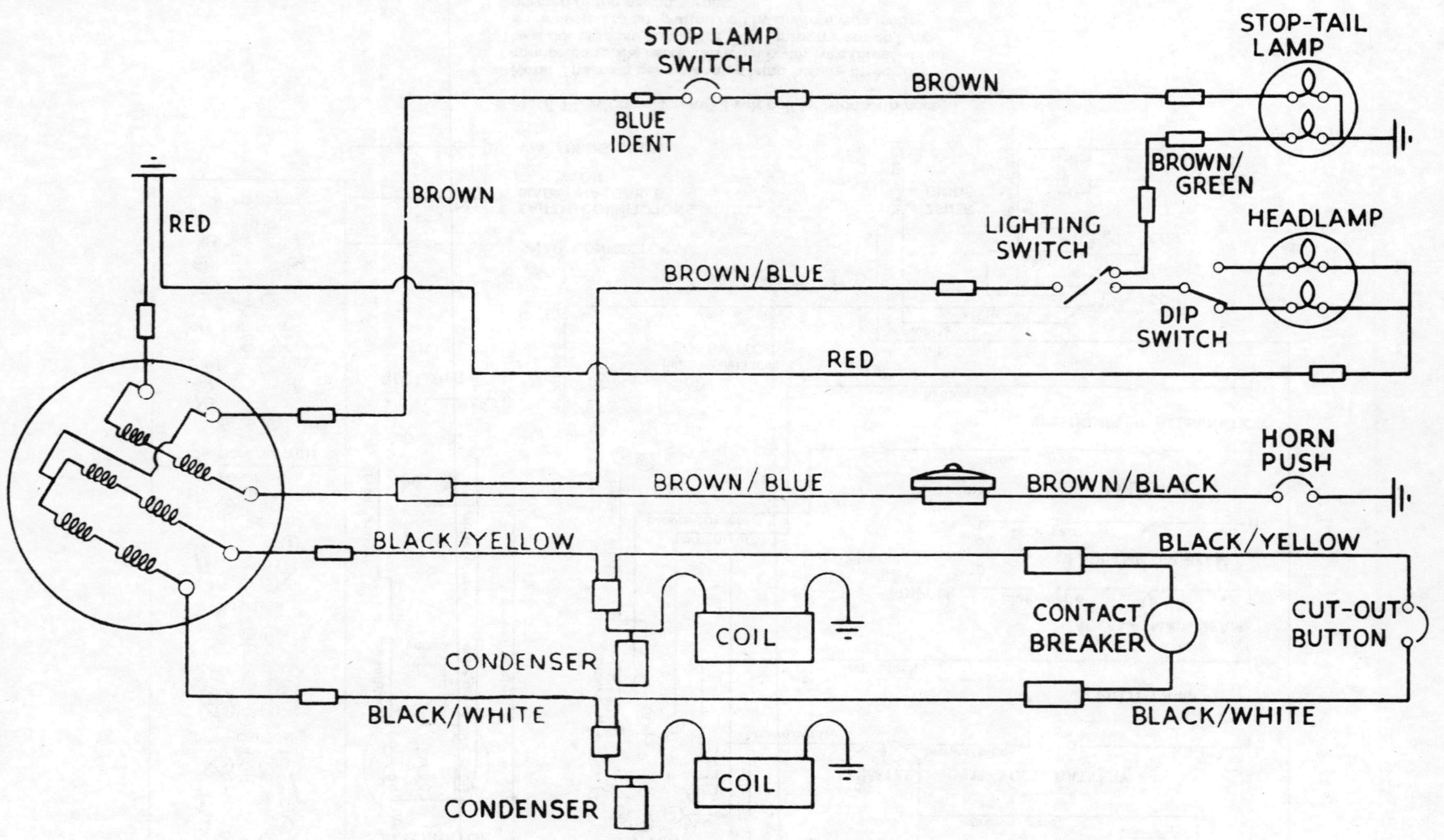

STOP-TAIL LAMP

HEADLAMP

BROWN/GREEN

DIP SWITCH

LIGHTING SWITCH

BROWN

STOP LAMP SWITCH

BLUE IDENT

BROWN/BLUE

RED

BROWN

RED

HORN PUSH

BROWN/BLACK

BROWN / BLUE

BLACK/YELLOW

CUT-OUT BUTTON

CONTACT BREAKER

BLACK/WHITE

COIL

CONDENSER

COIL

CONDENSER

BLACK/YELLOW

BLACK/WHITE

Fig. 8.15. Wiring diagram AC ignition (ET ignition)

Note: The main beam warning lamp (where fitted) is connected to the headlamp main beam wire by a double snap connector. No ignition warning lamp is fitted

ALTERNATOR AND STATOR DETAILS—
SPECIFICATIONS AND OUTPUT FIGURES

MODELS	System voltage	Ignition type	Alternator type	Stator No.
3TA, 5TA, T90, T100	12 V.	Coil	RM.19	47162 After H57083
3TA, 5TA, T90, T100	6 V.			47204
3TA, 5TA (POLICE)	6 V.	Coil	RM19/20	47167
T90, T100 (E.T.)	6 V.	A.C. IGN	RM.19	47188

Stator number	System voltage	D.C. input to battery amp. @ 3,000 r.p.m.			Alternator Output minimum A.C. volts @ 3,000 r.p.m.			Stator coil details		
		Off	Pilot	Head	A	B	C	No of coils	Turns per coil	S.W.G.
47162	6 V.	2·75	2·0	2·0	4·0	6·5	8·5	6	140	22
	12 V.	2·0*	2·1*	1·5*						
		4·8†	3·8†	1·8†						
47164	6 V.	2·7	0·9	1·6	4·5	7·0	9·5	6	122	21
47167	6 V.	6·6‡	6·6‡	13·6‡	7·7	11·6	13·2	6	74	19
47188	6 V.	Not applicable			5·0	1·5	3·5	2	250	25
								2	98	20
								1	98	20
								1	98	21
47204	12V.						8·5	as 47162		

2, 2 = IGN. ; 1, 1 = LIGHTS

Coil Ignition Machines
A=Green/White and Green/Black
B=Green/White and Green/Yellow
C=Green/White and {Green/Black / Green/Yellow} connected

* Zener in Circuit
† Zener disconnected
‡ With Boost Switch in Circuit

Note: On machines fitted with two lead stator, only test C is applicable as leads are coloured green/white and green/yellow.

A.C. Ignition Machines
A=Red and Brown/Blue
B=Black/Yellow and Black/White
C=Black/Yellow and Brown

Metric conversion tables

Inches	Decimals	Millimetres	Millimetres to Inches		Inches to Millimetres	
			mm	Inches	Inches	mm
1/64	0.015625	0.3969	0.01	0.00039	0.001	0.0254
1/32	0.03125	0.7937	0.02	0.00079	0.002	0.0508
3/64	0.046875	1.1906	0.03	0.00118	0.003	0.0762
1/16	0.0625	1.5875	0.04	0.00157	0.004	0.1016
5/64	0.078125	1.9844	0.05	0.00197	0.005	0.1270
3/32	0.09375	2.3812	0.06	0.00236	0.006	0.1524
7/64	0.109375	2.7781	0.07	0.00276	0.007	0.1778
1/8	0.125	3.1750	0.08	0.00315	0.008	0.2032
9/64	0.140625	3.5719	0.09	0.00354	0.009	0.2286
5/32	0.15625	3.9687	0.1	0.00394	0.01	0.254
11/64	0.171875	4.3656	0.2	0.00787	0.02	0.508
3/16	0.1875	4.7625	0.3	0.01181	0.03	0.762
13/64	0.203125	5.1594	0.4	0.01575	0.04	1.016
7/32	0.21875	5.5562	0.5	0.01969	0.05	1.270
15/64	0.234375	5.9531	0.6	0.02362	0.06	1.524
1/4	0.25	6.3500	0.7	0.02756	0.07	1.778
17/64	0.265625	6.7469	0.8	0.03150	0.08	2.032
9/32	0.28125	7.1437	0.9	0.03543	0.09	2.286
19/64	0.296875	7.5406	1	0.03937	0.1	2.54
5, 16	0.3125	7.9375	2	0.07874	0.2	5.08
21/64	0.328125	8.3344	3	0.11811	0.3	7.62
11/32	0.34375	8.7312	4	0.15748	0.4	10.16
23/64	0.359375	9.1281	5	0.19685	0.5	12.70
3/8	0.375	9.5250	6	0.23622	0.6	15.24
25/64	0.390625	9.9219	7	0.27559	0.7	17.78
13/32	0.40625	10.3187	8	0.31496	0.8	20.32
27/64	0.421875	10.7156	9	0.35433	0.9	22.86
7/16	0.4375	11.1125	10	0.39370	1	25.4
29/64	0.453125	11.5094	11	0.43307	2	50.8
15/32	0.46875	11.9062	12	0.47244	3	76.2
31/64	0.484375	12.3031	13	0.51181	4	101.6
1/2	0.5	12.7000	14	0.55118	5	127.0
33/64	0.515625	13.0969	15	0.59055	6	152.4
17/32	0.53125	13.4937	16	0.62992	7	177.8
35/64	0.546875	13.8906	17	0.66929	8	203.2
9/16	0.5625	14.2875	18	0.70866	9	228.6
37/64	0.578125	14.6844	19	0.74803	10	254.0
19/32	0.59375	15.0812	20	0.78740	11	279.4
39/64	0.609375	15.4781	21	0.82677	12	304.8
5/8	0.625	15.8750	22	0.86614	13	330.2
41/64	0.640625	16.2719	23	0.90551	14	355.6
21/32	0.65625	16.6687	24	0.94488	15	381.0
43/64	0.671875	17.0656	25	0.98425	16	406.4
11/16	0.6875	17.4625	26	1.02362	17	431.8
45/64	0.703125	17.8594	27	1.06299	18	457.2
23/32	0.71875	18.2562	28	1.10236	19	482.6
47/64	0.734375	18.6531	29	1.14173	20	508.0
3/4	0.75	19.0500	30	1.18110	21	533.4
49/64	0.765625	19.4469	31	1.22047	22	558.8
25/32	0.78125	19.8437	32	1.25984	23	584.2
51/64	0.796875	20.2406	33	1.29921	24	609.6
13/16	0.8125	20.6375	34	1.33858	25	635.0
53/64	0.828125	21.0344	35	1.37795	26	660.4
27/32	0.84375	21.4312	36	1.41732	27	685.8
55/64	0.859375	21.8281	37	1.4567	28	711.2
7/8	0.875	22.2250	38	1.4961	29	736.6
57/64	0.890625	22.6219	39	1.5354	30	762.0
29/32	0.90625	23.0187	40	1.5748	31	787.4
59/64	0.921875	23.4156	41	1.6142	32	812.8
15/16	0.9375	23.8121	42	1.6535	33	838.2
61/64	0.953125	24.2094	43	1.6929	34	863.6
31/32	0.96875	24.6062	44	1.7323	35	889.0
63/64	0.984375	25.0031	45	1.7717	36	914.4

English/American terminology

Because this book has been written in England, British English component names, phrases and spellings have been used throughout. American English usage is quite often different and whereas normally no confusion should occur, a list of equivalent terminology is given below.

English	American	English	American
Air filter	Air cleaner	Number plate	License plate
Alignment (headlamp)	Aim	Output or layshaft	Countershaft
Allen screw/key	Socket screw/wrench	Panniers	Side cases
Anticlockwise	Counterclockwise	Paraffin	Kerosene
Bottom/top gear	Low/high gear	Petrol	Gasoline
Bottom/top yoke	Bottom/top triple clamp	Petrol/fuel tank	Gas tank
Bush	Bushing	Pinking	Pinging
Carburettor	Carburetor	Rear suspension unit	Rear shock absorber
Catch	Latch	Rocker cover	Valve cover
Circlip	Snap ring	Selector	Shifter
Clutch drum	Clutch housing	Self-locking pliers	Vise-grips
Dip switch	Dimmer switch	Side or parking lamp	Parking or auxiliary light
Disulphide	Disulfide	Side or prop stand	Kick stand
Dynamo	DC generator	Silencer	Muffler
Earth	Ground	Spanner	Wrench
End float	End play	Split pin	Cotter pin
Engineer's blue	Machinist's dye	Stanchion	Tube
Exhaust pipe	Header	Sulphuric	Sulfuric
Fault diagnosis	Trouble shooting	Sump	Oil pan
Float chamber	Float bowl	Swinging arm	Swingarm
Footrest	Footpeg	Tab washer	Lock washer
Fuel/petrol tap	Petcock	Top box	Trunk
Gaiter	Boot	Torch	Flashlight
Gearbox	Transmission	Two/four stroke	Two/four cycle
Gearchange	Shift	Tyre	Tire
Gudgeon pin	Wrist/piston pin	Valve collar	Valve retainer
Indicator	Turn signal	Valve collets	Valve cotters
Inlet	Intake	Vice	Vise
Input shaft or mainshaft	Mainshaft	Wheel spindle	Axle
Kickstart	Kickstarter	White spirit	Stoddard solvent
Lower leg	Slider	Windscreen	Windshield
Mudguard	Fender		

Index